# Making Miniatures With 3D Printers #2

SLIDING SCALE RECORDS

## Dedication

To my lovely wife Angie who insisted I write another book.

## Publishing Data

First edition published 2026 (SSFF02)

Text copyright Frank Fisher and Angie Scarr

Illustration, composites and photographs copyright Frank Fisher

Design by Frank Fisher and Angie Scarr

Plaza De Andalucía 1, Campofrío, 21668, Huelva, Spain.

## ISBN 978-84-126023-9-5

# Contents

# Foreword by Angie Scarr

## 'Learning by doing'

If you're anything at all like me your first thought might be "That's tech ... I can't do that". Fortunately I have always had Frank to help me when I need to turn design ideas into reality. But what about people who don't have a tame computer geek? A couple of years ago I said to Frank "Hey, what about writing a book as if you were writing it for me. So that people like me, who really want to come up with their own designs for 3D printers can cut through all the scary technical stuff and learn to do it bit-by-bit?"

So that is where our first 3D printing book "**Making Miniatures With 3D Printers** *Designing for dollhouse, book nooks and dioramas*" came from. This second in the series starts off simply and reaches more rapidly into some of the more complicated techniques. But we have hidden the techie stuff at the back so as not to frighten newcomers to the hobby.

If you'd like to take things slowly we do recommend the first book for more of the early simple steps and more of the problems and solutions. I hope that both lead you to be a bit more fearless, through the technical stuff that scares me and might intimidate you. It's written for people like us! I haven't had to do everything in this book to prove it can be done, but I did do the stool to make absolutely sure it worked. If you have never designed for a 3D printer before, we both recommend that you do the first project yourself before trying anything else. Actually even if you have, doing the stool will help you get into Frank's style of teaching. This book should help you to imagine how you can create your own files and of course adapt bought files.

I hope whatever level you want to reach, you have some fun with the learning process and don't expect that everything will work perfectly every time. It doesn't. Being an expert comes from making a lot of mistakes! 3D printing is still a relatively new hobby, and machines sometimes need a manufacturer fix too, so getting to know your machine and joining groups of other interested makers is always a good idea. I hope this book gives you the confidence and the immense satisfaction to be had from knowing that at least part of your design is all yours!

In this book we add examples of how useful multicolour printers can be, but the projects can be used on your older machines too. The FAQs at the end reflect the most common questions Frank has had to answer, but doesn't repeat ones from the first book which are still regularly asked!

## A few words on language used in this book

Frank and I did disagree about what language we used for dimensions and in 3D printing design. The standard is in axes X, Y and Z with

**X being the width (left to right)**

**Y being front to back (depth)**

**Z being bottom to top (height) of the object as it is being designed.**

However this can get very confusing when designing or printing something in another orientation to the one which we're accustomed to seeing it or if it gets spun. For design purposes when Frank is designing for something printed on the bed in an unusual orientation he will still use height meaning the height off the print bed, which may or may not be the height of the free-standing object after printing, depending on whether it is being designed lying on its back as is often the case, or standing up. Some users might think of this as the 'depth' of the print.

You might find it helpful to label your designs, if drawn on paper, in whatever way is comfortable for you.

# Introductory Project: Kitchen Stool

## This project will:

- *Introduce Tinkercad*
- *Show how to use pre-made shapes to produce simple designs*

If you haven't used Tinkercad before and haven't tried the first book, here is an easy introduction. We will make a stool in a few minutes with a free Tinkercad account. If you haven't created one yet go to www.tinkercad.com and sign up. You can use e.g. Google login to create an account here quite quickly.

If in doubt when asked what kind of account you want choose Personal.

You should then see a page with a "Create" button at the right - click it. These pages do change over time however as the marketing folks change things around.

Now you should see a blue grid with Workspace written on it and tools at top, left and right. We are ready to create.

Starting with the legs click on the cone at the right hand side and drag it onto the workspace.

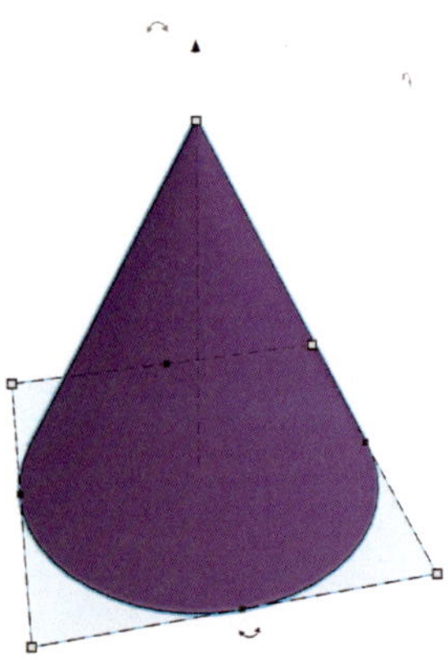

You will see it has 5 white boxes and a black cone. The corner boxes let you type in a measurement so click one and there will be the number 20, click on it so it turns blue then type 3.5 There will be another number 20 click and type 3.5 there also.

The white box at the top lets you type a height, so type 200 over the 20 there.

If at any time the object is too big for the screen click the Fit to All button on the left hand side bar. If your view looks different to mine don't worry as long as you have the long thin cone.

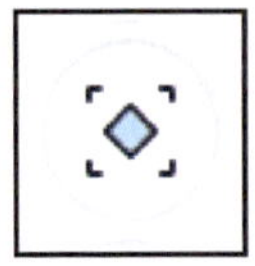

If you want to see it from a different angle you can click the View Cube and drag it around.

If you totally lose the objects then just click the Home button and it will take you back to an overview of the workspace.

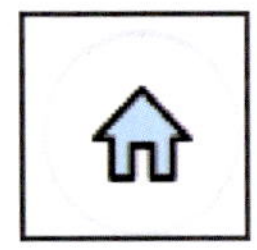

We now have a long, thin cone. In some versions we could directly tell Tinkercad how thick we want it at the end rather than a point but instead we will learn a technique to adapt any shape.

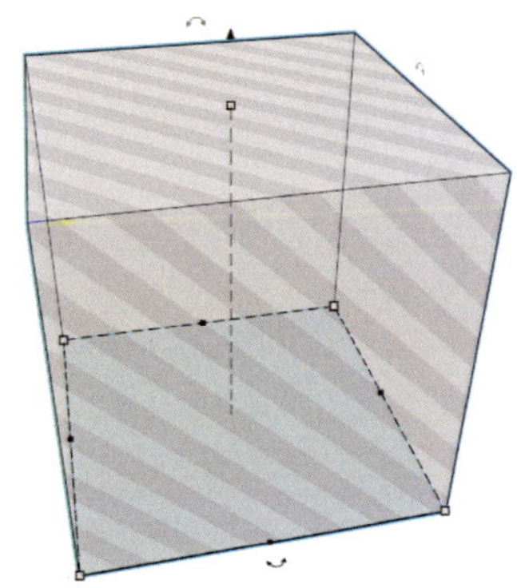

The grey cube makes holes, a bit like erasing, so drag one of those onto the workspace and click on the top white box. Then type 134 over the number 20.

We want to move this to the top of the leg so that we can cut it off and leave a flat surface at the top rather than a point (which wouldn't print very well or be stable to stand on).

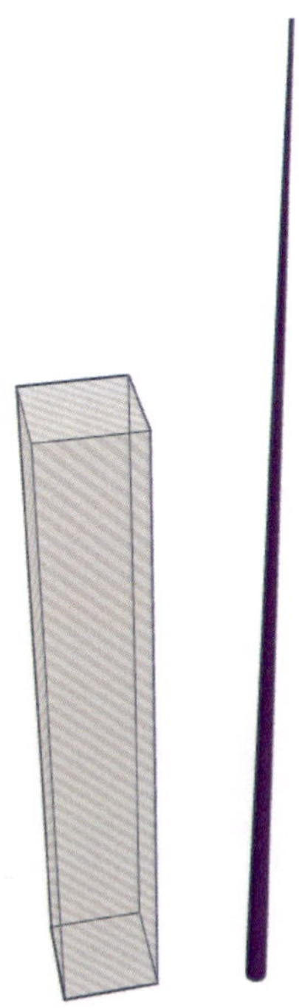

Select the leg and the hole by dragging the mouse across them - you will see a dotted red box that disappears when you let go.

At the top left is the Align tool, click it and click the two centre circles at the front to back and left to right on the workspace. You can see that I have highlighted them with a red circle. At the side you will see the height options also circled in red, click on the top circle.

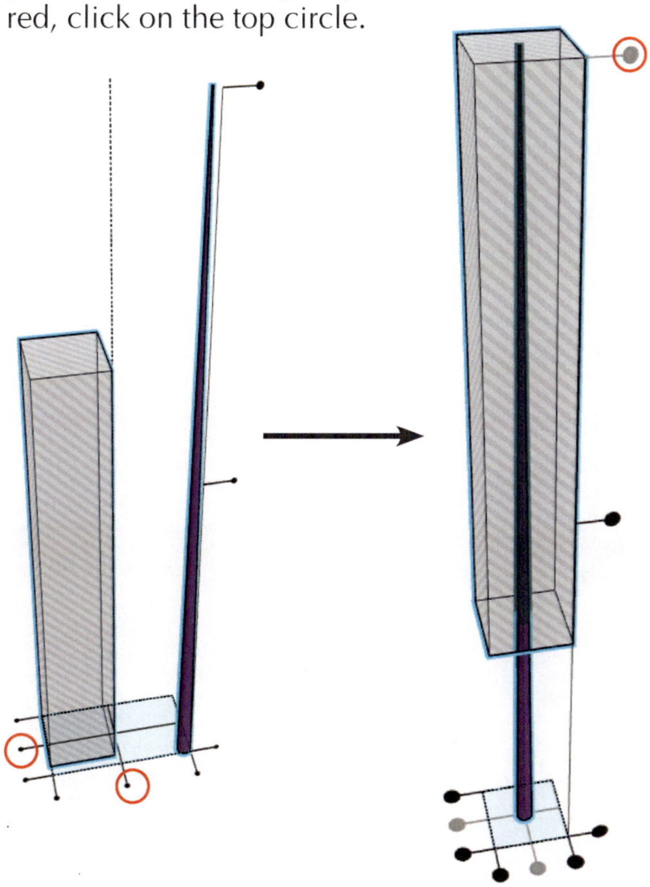

Now click the Union button at the top right. It will take the hole away from the top of the cone leaving a gently sloping leg with a flat top.

You might need to zoom in again at this point using Fit To All.

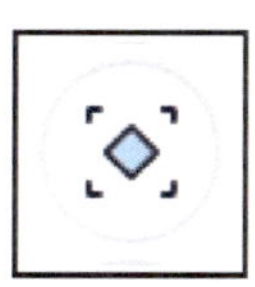

**Congratulations**, you made your first 3D object.

If you click on the leg now you will see two curved arrows at the top and one at the bottom. One might be side on to you and hard to see. At this point it is also likely to be a little too zoomed in, so turn the mouse wheel towards you to shrink the view or drag the leg lower down the screen.

Up until here any mistakes could be easily rectified by simply changing the numbers again, however once we rotate only the Undo button at top left can save you.

Click on one of the arrows at the top, hold and drag the mouse, without letting go, on the top outside edge of the wheel. Drag to to get "-7". This might be left or right depending on your

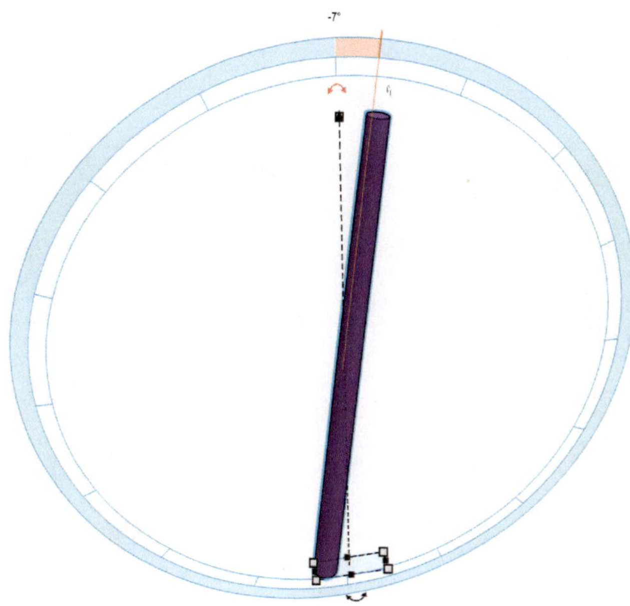

viewing angle. You can type this if it is a little fiddly. Do the same with the other top arrow. It might appear as an indistinct mark if it is side on to you, so you may need to move the View Cube to see it. This angles the leg slightly outwards front to back and left to right.

This last part is the easiest place to make mistakes that you won't see until you position the leg. Remember the undo button.

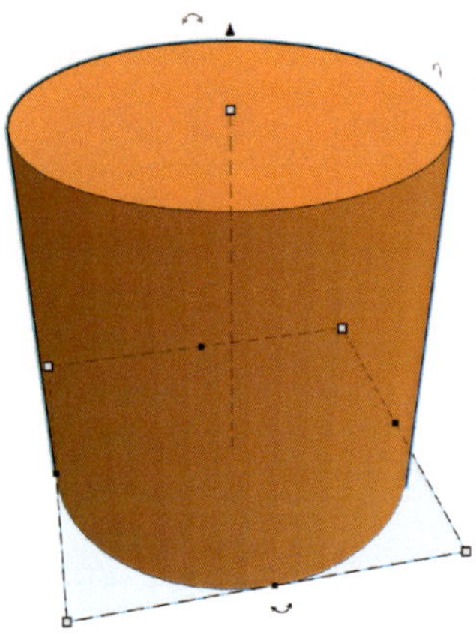

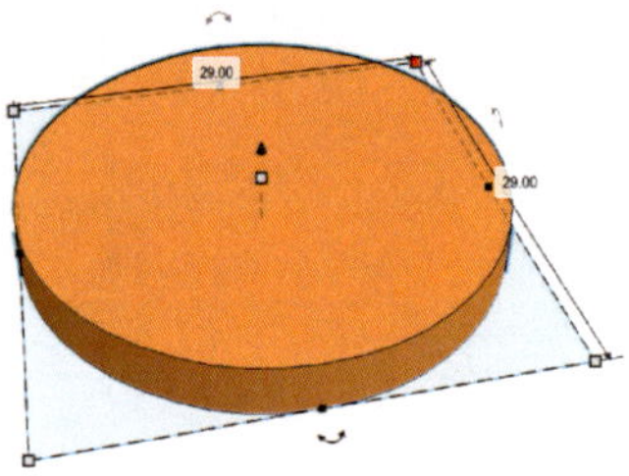

Now we will make a seat for the stool. Drag an orange cylinder onto the workspace.

By default all measurements are 20, but click on the white boxes and make the sides 29. Click on the top box and make it 3.

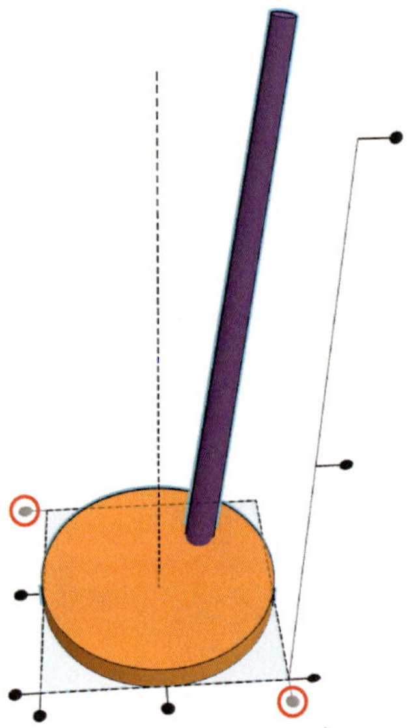

Select the seat and the leg by dragging acrosss them and click Align. Click on the back circle and the right circle. If the leg appears on the edge of the seat instead of inside, select and align it to a different corner until it fits.

If it still doesn't fit you may have made a mistake on the angle and need to go back. Either use Undo or recreate the leg.

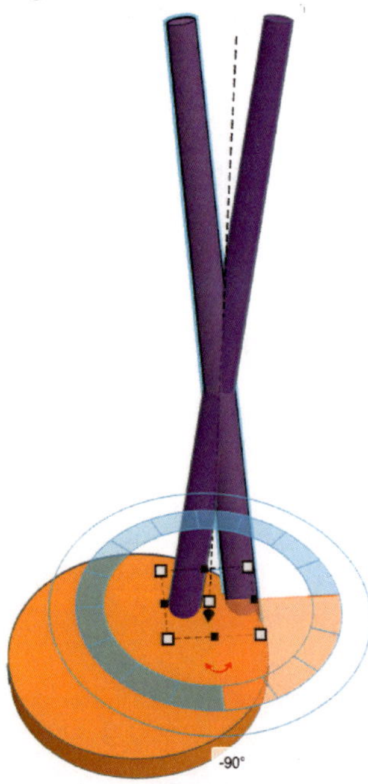



Click off then click on the leg. Press the Duplicate and Repeat button. You won't see the copy yet. Now use the curved arrow at the bottom to spin it 90 degrees - if you go inside the circle it moves in easy jumps like 45, 90 etc.

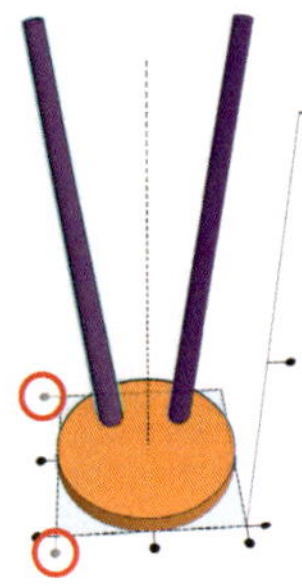

Hold the shift key on your keyboard (it may be an up arrow) and click on the stool base, now Align it to the back and the left.

Select both the legs and Duplicate them.

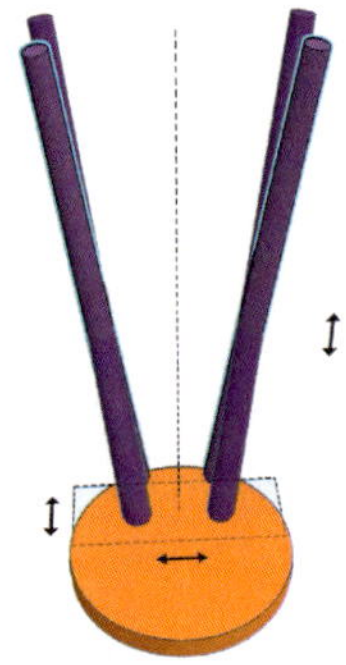

Now click the Mirror button and click the front to back arrow.

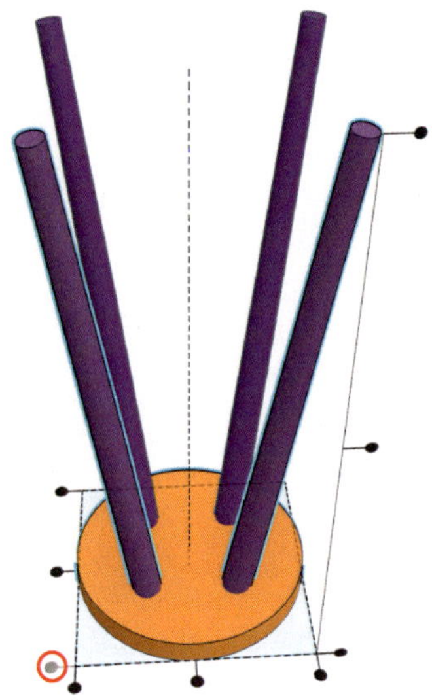

Shift click on the seat and Align them to the front. Select the four legs and the seat and click Bundle.

Finally, because we angled the legs, they would be a little wobbly when standing as the ends are no longer flat.

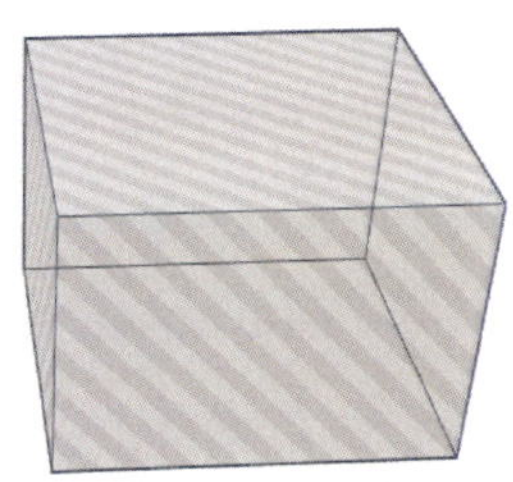

Drag a hole cube onto the workspace and make it 30 mm by 30 mm at the base.

Drag the cone on top and raise it to 64 mm high.

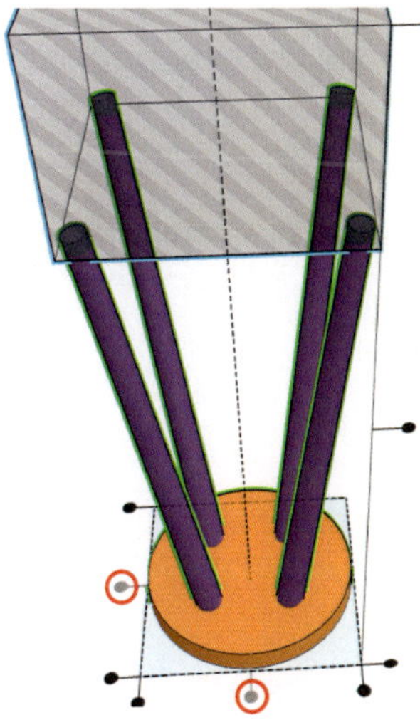

Select the stool and hole and align them to centre at the base. Now click Union. Note that at any time there may be delays on a web service like this - an action may be instant or take a few seconds.

This union has shaved the angled top off the legs to make them perfectly level.

This gives a plain stool but I would like to add a ring part way up for resting the feet as it is quite tall.

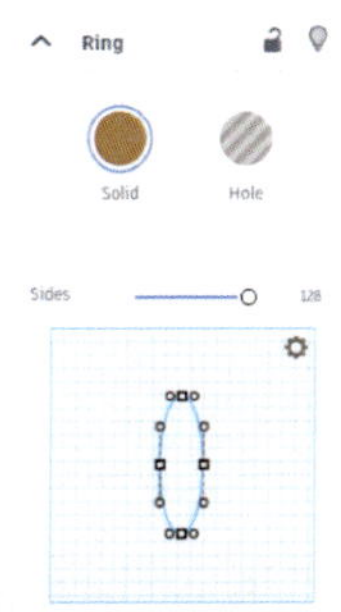


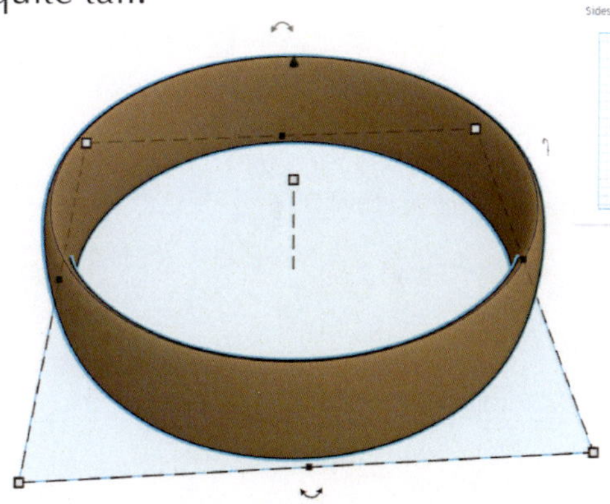

Drag a Ring onto the workspace. Drag the Sides slider to full (128 normally). This smoothes out the shape.

Click on the white boxes to make it 25 mm by 25 mm and 2 mm high.

Click Export (top right), then STL and name it "stool ring".

Click on the stool, Click Export, then STL and name it stool body.

I have taken them separately as the ring is quite fine and may need help printing with some settings in the slicer later (the slicer is the software that takes the STL files you just saved and gets them ready for the 3D printer.

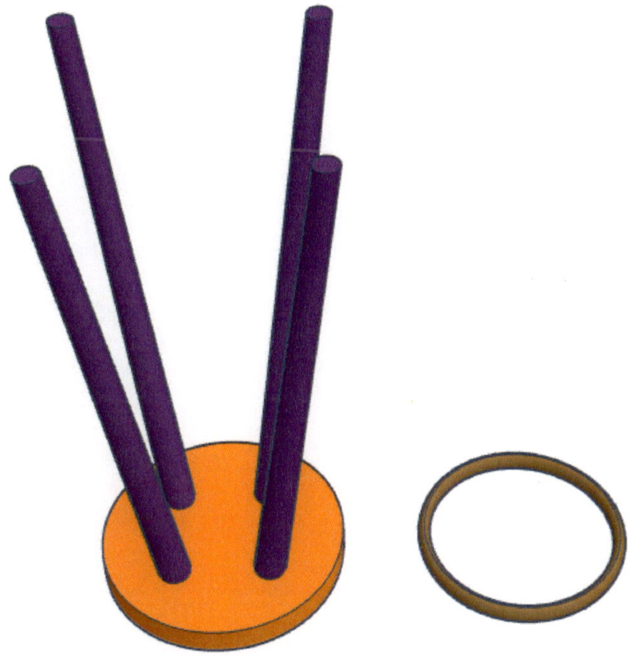

If you want to see what a slicer does see the appendix).

This ring is designed to be glued insde the legs. If you want to make a stronger ring you can make it from an inverted cone but that is a higher level of complexity.

When you slice the ring you will need to enable Brim on filament printers (at the top right near Infill % on Prusaslicer) as it puts a fine layer around the ring to help it stick to the bed during printing. The brim is very easy to remove.

If you want to start a new project at any time your current work will be saved in your gallery.

Click this button up at the top left near the Tinkercad logo and select New Design to start clean or choose one of your saved projects.

Your projects are always given a random name by Tinkercad but you can click on the name (next to this button) and type whatever you want.

# Simple Fire Surround

## This project teaches you how to:

- *Use Tinkercad and Inkscape together*
- *Use Inkscape to map out component shapes*
- *Work with curves*
- *Make recesses*

# Simple Fire Surround

As we begin the main projects in this book it is a good idea to get into the habit of re-colouring the individual parts so you don't lose your way and confuse pieces. I show you how I colour them. In the screenshots I colour each cube differently to more easily see the pieces. The colour at this stage makes no difference and will have no effect on the print, it's just to make designing simpler. Also it can be helpful to name projects you have created, both when leaving a project half done and also so that you can create a bank of your personally designed elements to re-use in later projects.

There is a cheat sheet for using Tinkercad in the back of this book to help find the buttons and commands we will use.

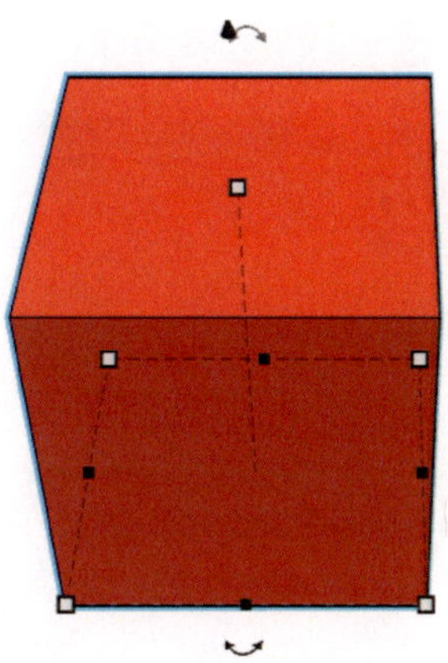

Start by adding a cube 'primitive' (primitives are all the ready made shapes at the right of the screen), it is red on the basic shapes dropdown. Click on it, click the white corner box and make it 14.5 mm wide x 99 mm either by dragging or typing the number. Then click the square top centre and make it 10 mm. This will be the right hand side leg. I have coloured it pink - use

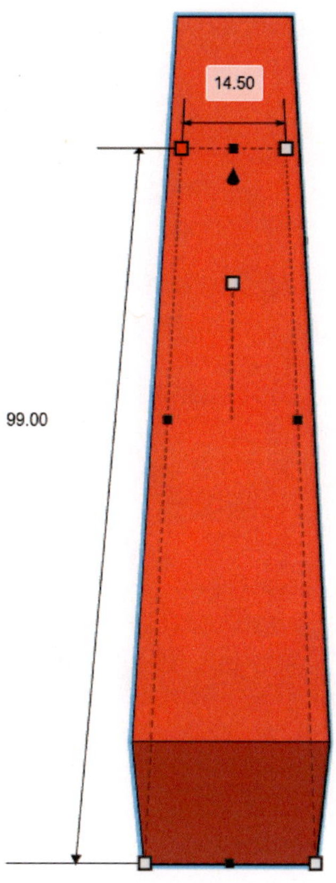


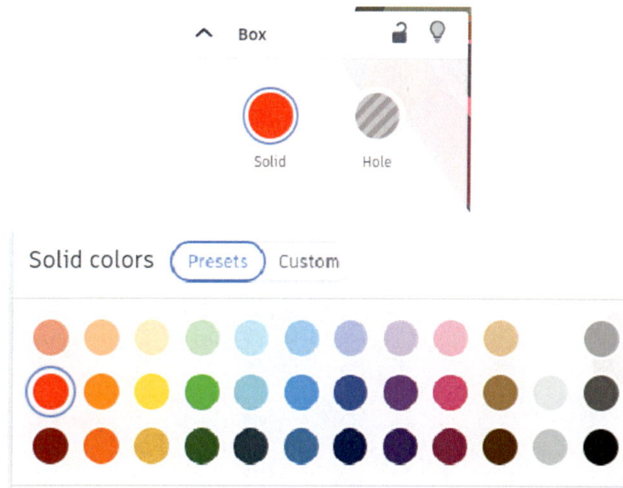


the colour selection at the top right and choose pink from the presets.

Next we will add a mantel. Place a cube and make it 10mm x 135 mm wide x 21 mm tall. Just roughly place it on to the top of the leg for now, we will do the exact alignment when we

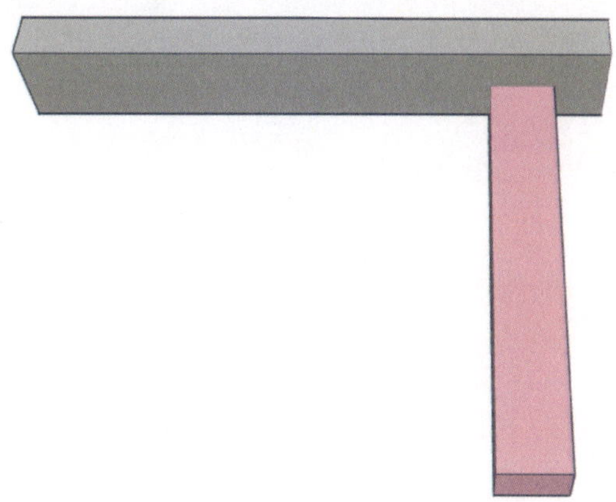

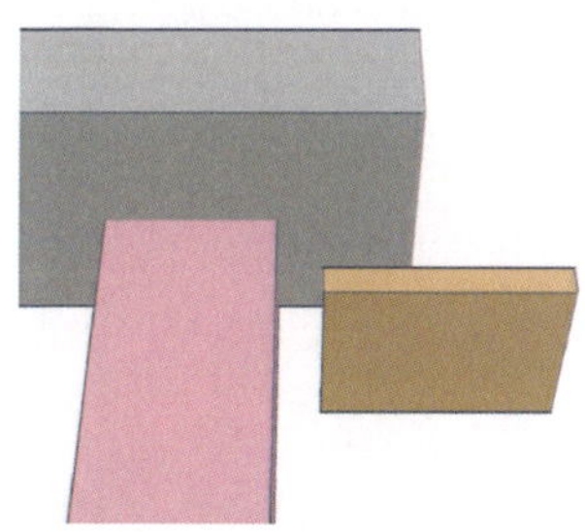

have the final pieces. I have coloured it grey.

Now we will add some simple decoration using more cubes. Add a cube and make it 20 mm wide x 2.5 mm x 13mm high. I have coloured it light brown. Repeat it using this button at the top left and move the copy near the bottom of the leg.

To make a plinth for the bottom of the leg, place another cube and make it 18 mm x 18 mm and 11.5mm high. I have coloured it blue.

I am going to add a curve at the top to look like a decorative support. Place a cylinder and make it 20 mm wide by 29 mm, it should already be 10 mm high. Place a cube and make it 6 mm wide by 14.5 mm, again it should be 10 mm high. Position it so that it intersects the cylinder leaving a small amount top and bottom.

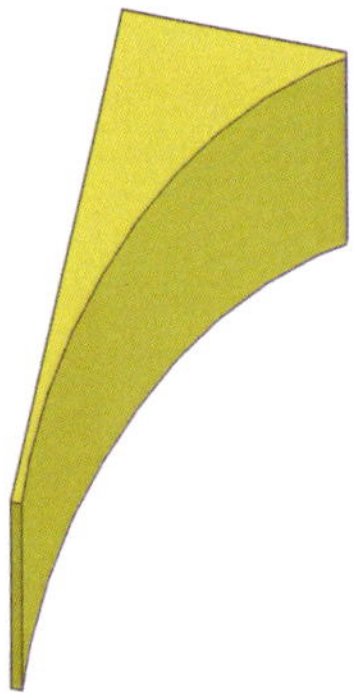

If you need to, use the arrow keys and change the Snap Grid (bottom left) to 0.1 mm.

I have coloured the curved piece yellow.

A final piece of decoration will be a recess in the leg. You can make as much or as little decoration as you like but I think for this example a simple groove is enough. Place a cube and make it 10 mm wide by 24 mm and 2 mm high. Place a cylinder and make it 10 mm by 10 mm and 2 mm high. Position it so its midpoint is aligned with the top of the cube, then repeat the cylinder by clicking this button and move it to the bottom. We can now select all 3 and "union" them by clicking this button.

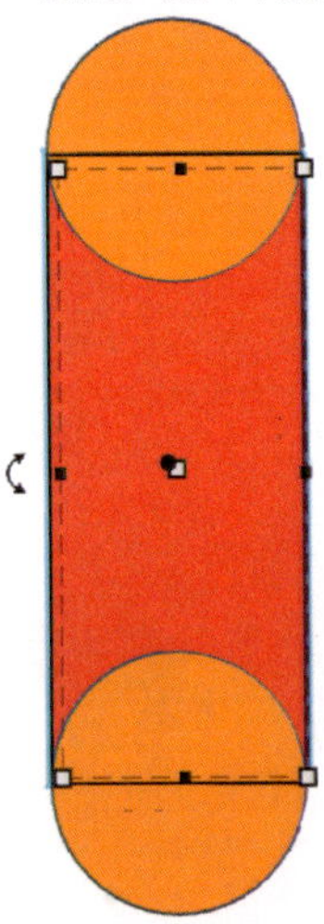

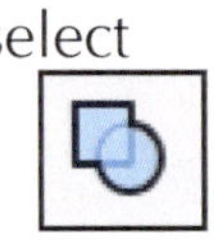

Selection can be by holding the mouse button down and dragging the mouse across the objects you want, or by clicking one then holding SHIFT on the keyboard and clicking the others. Turn them from a solid into a hole.

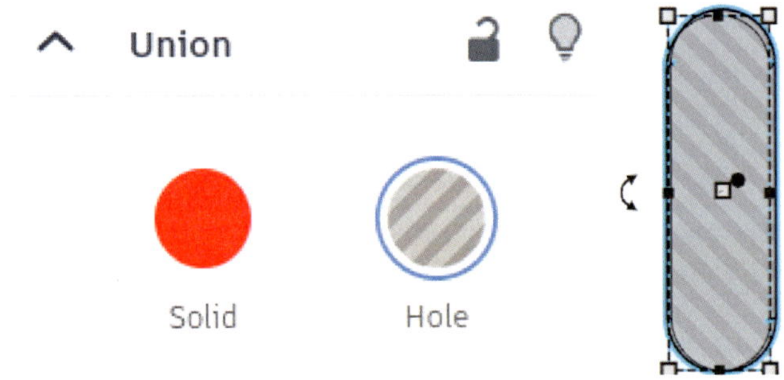


We now have all the pieces and can do a final alignment with them to make the left leg and mantel.

Start by aligning the plinth to the bottom of the leg. Select both and click the align button. Now three sets of black circles appear: left to right, top to bottom and up and down. Align the plinth bottom and centre, both should already be at down. When the objects are aligned to a direction it will grey out. You can see all 3 have a greyed option.

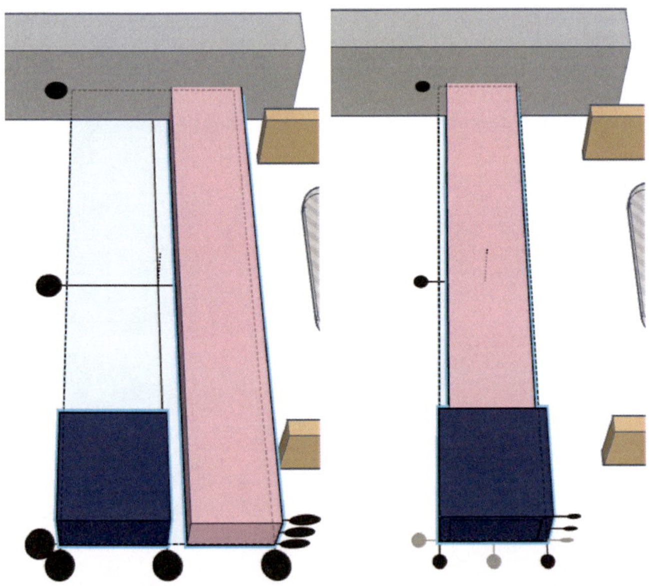

Next repeat with the decorative brown squares. Don't worry about exactly where they are just yet, as long as they are centred on the leg. You must also have the plinth selected when you align as it isn't attached yet and the leg will move away from it if you don't. Now add the decorative hole with shift-click. Align the hole to top, centre, centre. You can deselect all and move the hole with the arrow keys until it is here you like it.

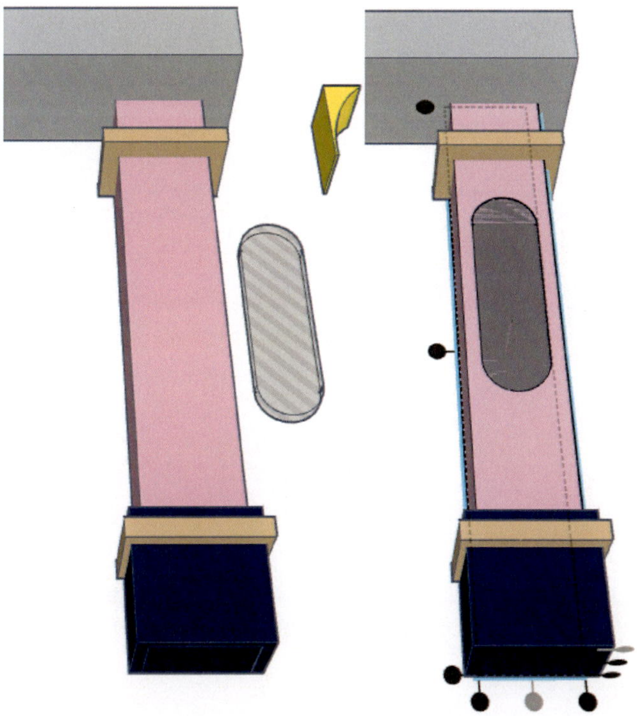

Select all the leg parts and union them  which will show you how the object appears when printed with the hole carved into it. Align the leg to the top of the mantle. At this point it is unlikely everything on the leg will be in the final position but don't worry.

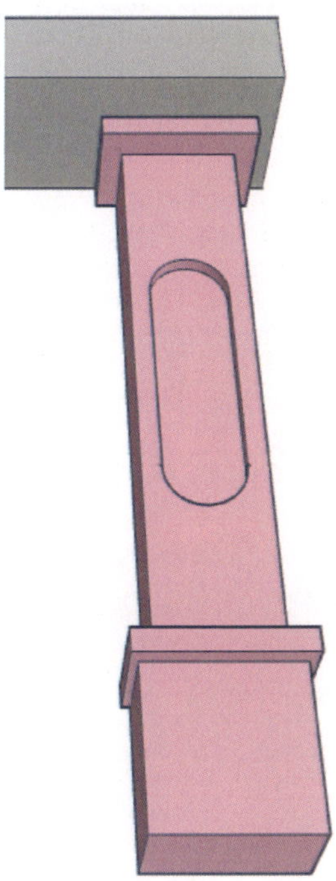

We are now ready to align the curved decoration to the right edge of the mantel. Use the arrow keys to make it perfectly fit the bottom of the mantel. Now we can arrow key the leg along until it is exactly touching the curved piece.

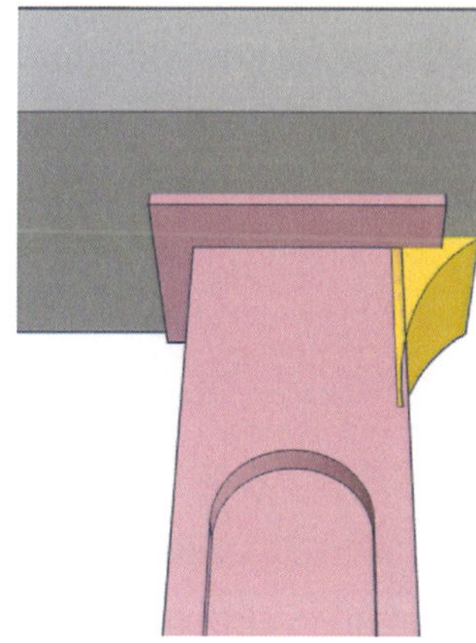

For the final stage of alignment unbundle the leg with this button  and move the brown cubes up or down with the arrow key until they are perfect too. It is important not to use the mouse as you already have it aligned in the centre which you can lose with the free movement of the mouse.

Finally copy the left leg with repeat  and move it to the left, then flip it horizontally to mirror it with this button  and align it to the mantel top and left.

Select the two legs and mantel and click union. I have coloured the result brown. This can be sent off to the printer at any point. Congratulations, you made a 3D model!

# Decorative Inner Trim

Now to partly fill the space and make a complete fireplace. I made a decorative tile effect to simulate cast iron. Make a cube of 106 mm wide by 96 mm and 1.5mm high. I coloured it black.

Make a hole bottom centre 58 mm wide by 47 mm and union them to make an opening.

Now make a pyramid 10.8 mm by 10.8 mm and 15 mm high.

Next make a hole cube 20 mm x 20 mm and 12.75 mm high. Align it to the top of the pyramid and union them. We now have a slightly sloped tile.

Make 7 copies and spread them across to fill approximately 93 mm. Bundle them all by selecting and clicking this button. This is just a kind of grouping and

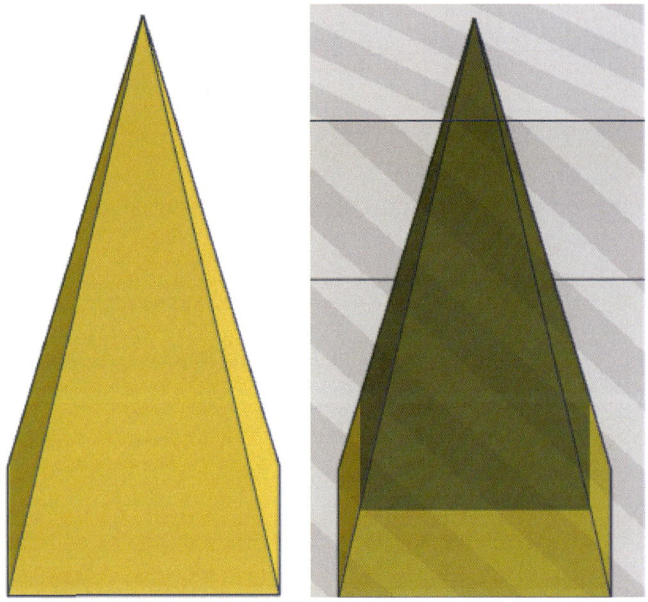

doesn't weld holes into solids, it just carries everything in place.

I have chosen to have 2 rows like this then make the tiles smaller for the vertical section. Any spaces I filled with long tiles.

You can see I have left a border (in black) to glue to the back of the fire surround.

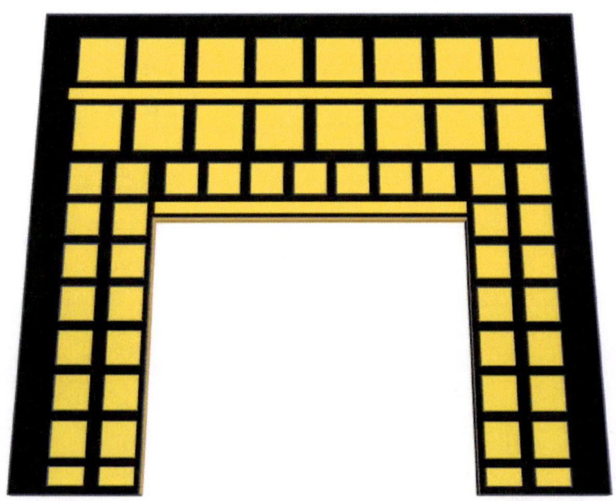

## Firebox

Next we line the opening of the chimney with the firebox. It is quite simple to make the box at the back, but we will also slope and close the top into a tube.

Make a cube of 69.5 mm wide by 26.25 mm and 56.5 mm high.

Make a hole cube of 59.5 mm wide by 24.25 and 62 mm high. Move this cube off the baseline using the black cone at the top of the cube, pull it up (it will turn red as you do) until it reads 2.00 at the bottom right.

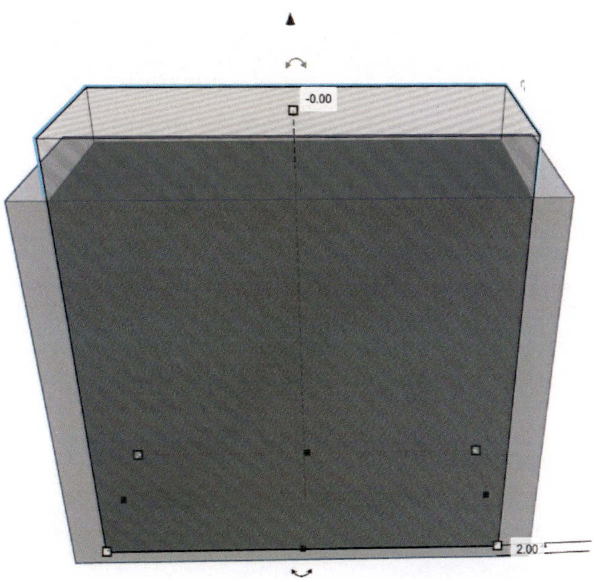


You can now select both cubes and union them. That's the simple part done, now we need to close it in at the top.

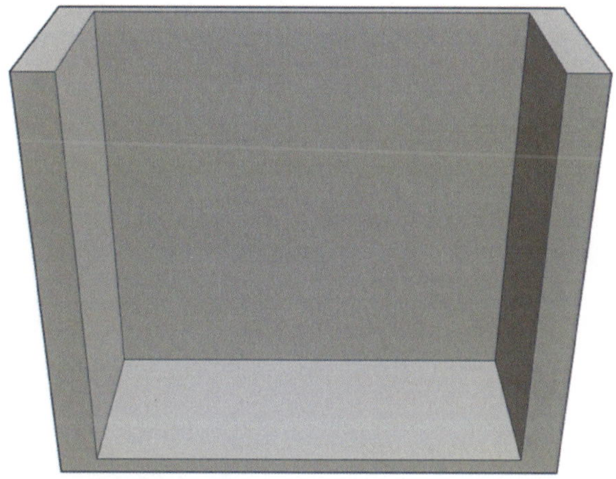

## Smoke Chamber

We are going to start the sloped top and tube with a pyramid. Make it 69.4 mm wide by 24.25 mm and 60 mm high. Make a hole cube 40 mm wide by 20 mm and 20 mm high, align it top centre to the pyramid and union them.

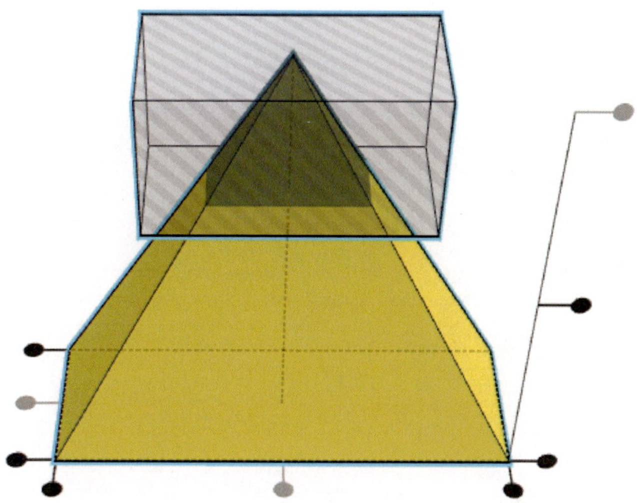

Next we will make a tube at the top. Make a cylinder 18.5 mm by 18.5 mm and 29 mm high. Raise it to 20 mm above baseline.

Make a hole cylinder 15.25 mm by 15.25 mm and 29 mm high. Also raise this to 20 mm off baseline.

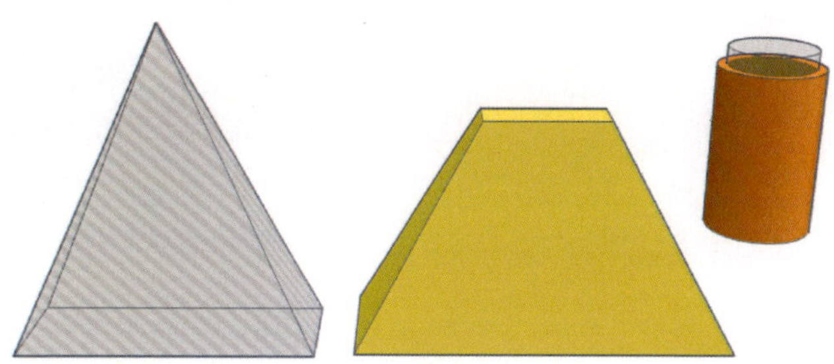

Now make a hole pyramid 55 mm wide by 20 mm and 57.5 mm high. Align all the pieces to the centre of the chopped pyramid.

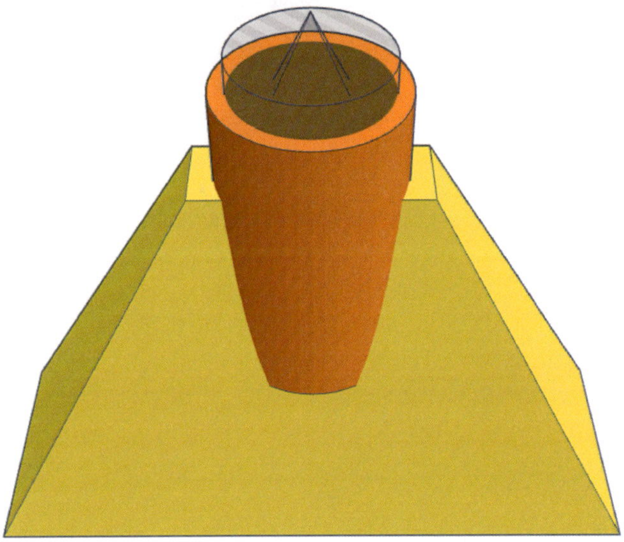

Union them to make the finished piece.

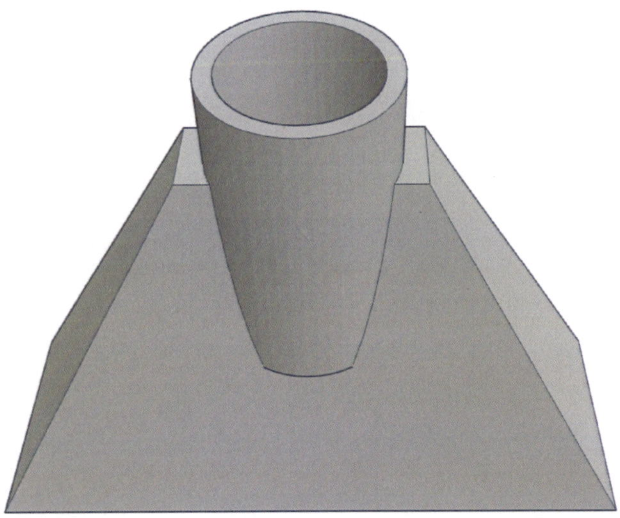

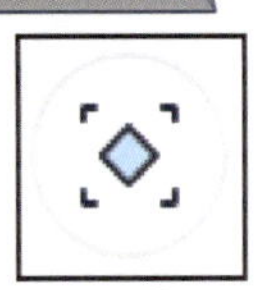

Have a look around it first by clicking the Fit View To Selected button to centre the viewpoint on the item.

Now you can spin the View Cube by dragging it with the mouse.

# Fire Basket

*Introducing Inkscape*

Maybe we could make this straight in Tinkercad but it would involve a lot of messing around to get the exact shapes shown here. I find it easier to use something like Inkscape (used here), CorelDraw or Illustrator - vector graphics programs that make SVG files.

> *I use Inkscape for this example - it is also free and can be used on almost any desktop machine.*

I have drawn a rectangle 47.5 mm wide by 12.5 mm. In Inkscape then click Path -> Object To Path so we can edit it easily. Draw another box over it that is 42 mm wide as a guide.

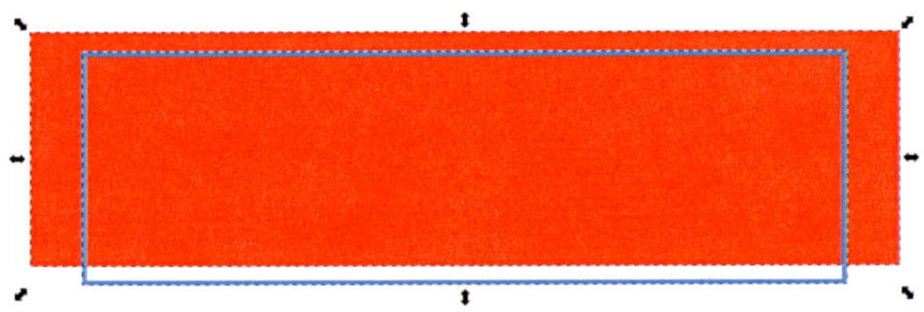

Centre the two boxes by selecting both and using align and clicking the "centre on vertical" button.

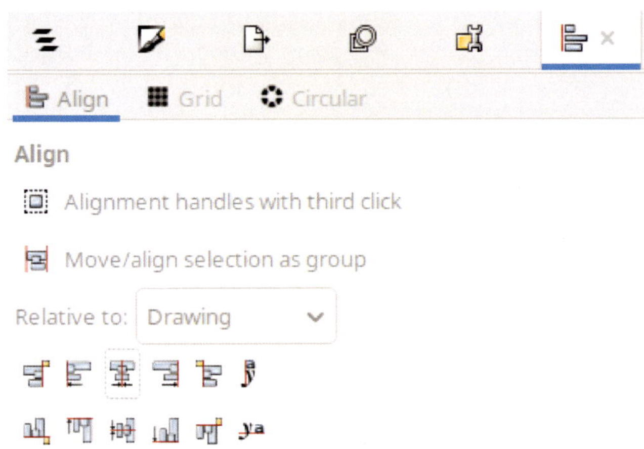

Now click the node tool and click on the main rectangle then on the bottom left node (small white circle). Use the arrow key to move it in towards the guide box, then repeat at the right hand side. It doesn't have to be exact as long as it is symmetrical.

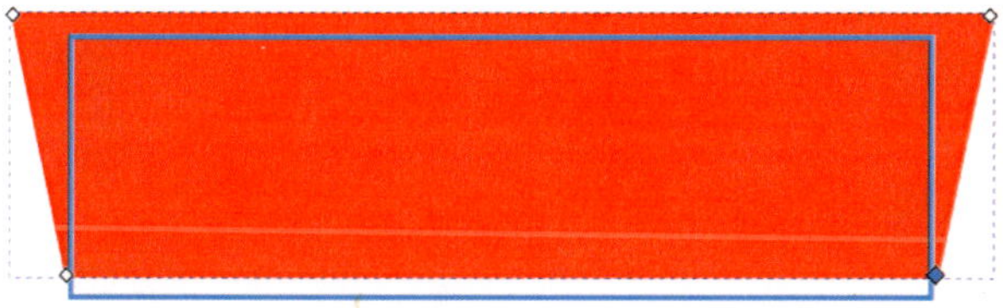

Copy and paste the main rectangle to use later for the front piece. Move the copy away for now.

Draw a circle and make it 48.75 mm by 31.25 mm. Align it "vertical centre" and "horizontal bottom" to the first shape.

Make a circle 15.25 mm wide by 17 mm, align it vertically centre to the other two objects, then click off and onto the new small circle and use the arrow keys to move it up so about half is showing above the circle.

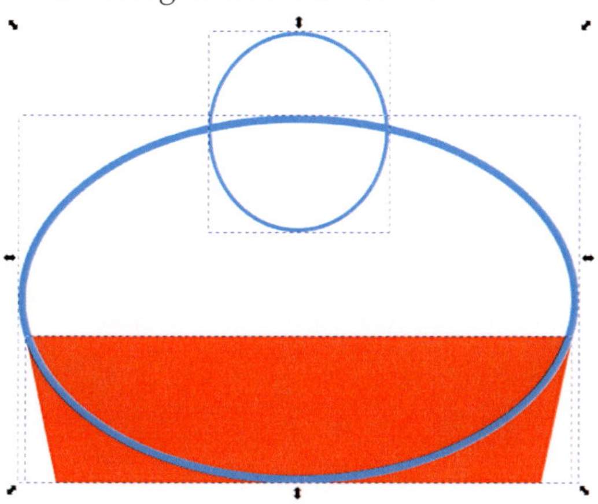

Select all three and click Path -> Union. We now have the back plate. You can leave it as is or play with the node tool to change the shape a little. Or start again from the front piece to make your own design, just make sure it fits the first piece as this will become a box later.

For the front piece we want to make a grill, so bring back the piece we made earlier and draw a box of 2 mm wide by 8mm. It is easier to click on a white colour from the bottom swatch at this point. Place it near the bottom edge.

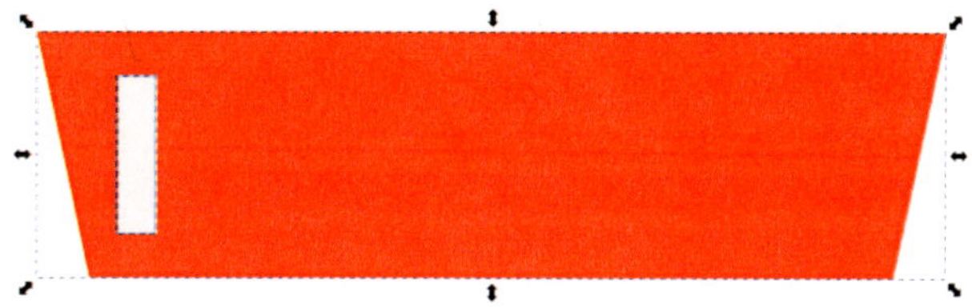

Now click Edit -> Duplicate and make 9 copies of it. You may see a shortcut there Ctrl-D which you can also use for speed. Use the arrow key to move one of the boxes near the right hand bottom edge. Now select all the boxes and use Align, Distribute and the middle button "Distribute horizontally with even spacing".

So we can align these perfectly against the initial piece click Object -> Group. Then select the initial piece too and align both horizontal and vertical centres.

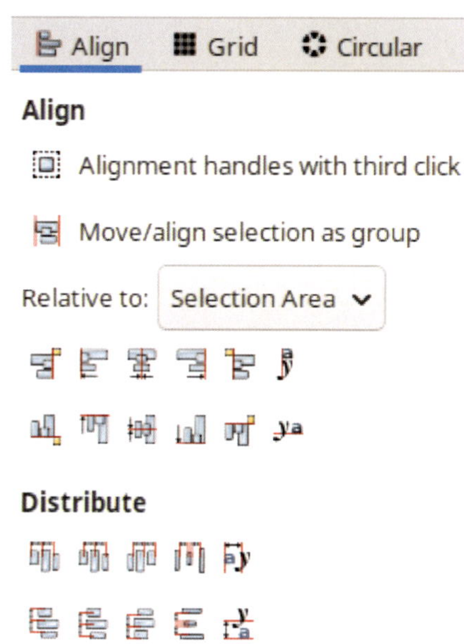

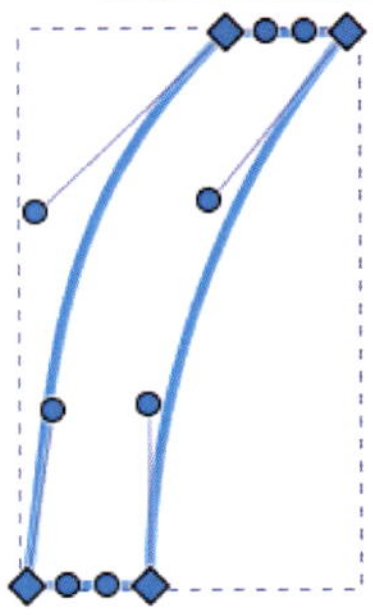

For the feet I made a square 3 mm wide by 13 mm. To turn it into a curve first click Path -> Object To Path. Click the node tool and select the top nodes (the diamonds at the corners). Use the arrow key to move it across about 8 times. This now gives and angled leg. But we can do better and add curves to it.

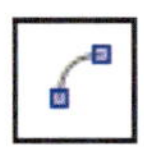

Select all the nodes this time and click the Add Curve Handles button. Pull the newly added handles until you are happy with the shape. We now have a curved foot, a front grill and a back piece to take out to Tinkercad as SVG files. Click each in turn, File -> Export and fill out the filename for each part so that you can find them later.

Click "Export Selected Only" otherwise your whole drawing is exported. Click "Plain SVG" also to maximise compatibility with other software.

## Assembly

Now we move into Tinkercad to assemble the pieces we just created.

Import the back piece and make it 2 mm high (I coloured it pink).

Import the grate and make it 1.5 mm high.

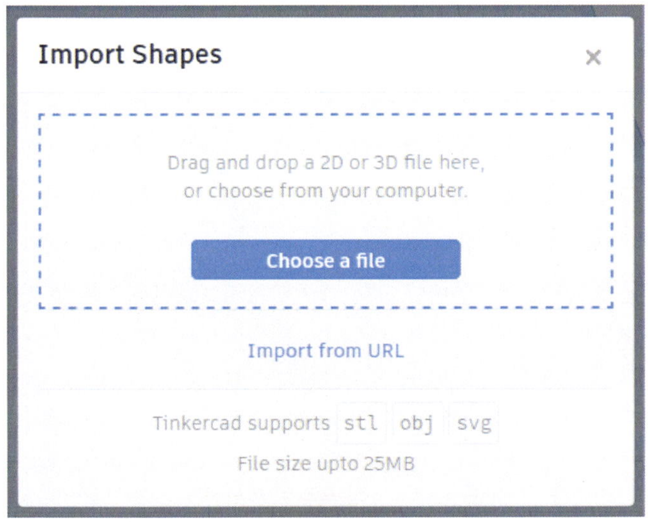


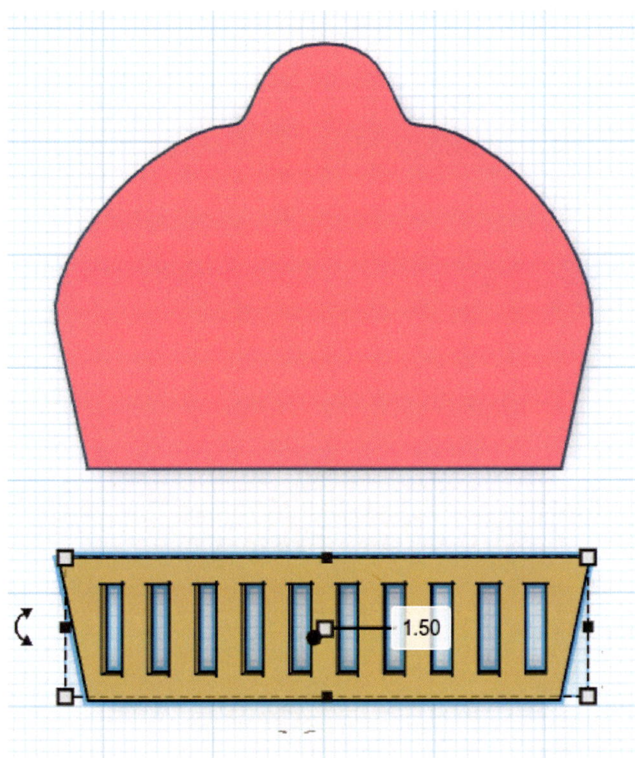


Import the leg and make it 4 mm high.

Copy and mirror the leg horizontally.

Make a cube 42 mm wide by 1.5 mm and 20 mm high for the bottom of the basket (I coloured it green).

Make a cube 40 mm wide by 3 mm and 4 mm high for the leg brace.

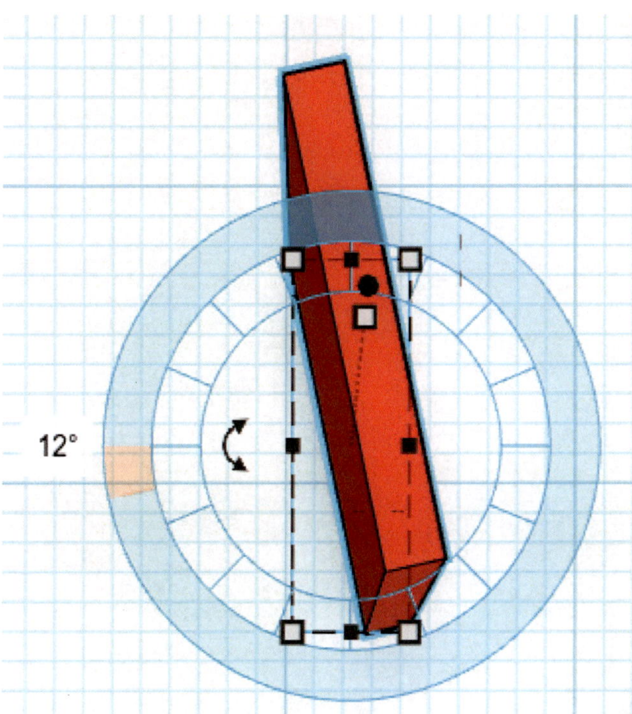


Make a cube 1.5 mm wide by 12.5 mm and 20 mm high for the side of the basket. Click on it and use drag rotation arrows to turn it 12 degrees. If you are inside the rotation circle it goes in steps like 45, 90 etc. so you will need to be outside to be able to go 1 degree at a time. Or just click in the box showing the angle and type 12 if you find it difficult.

Align to bottom and left of the pink back piece. Repeat the side piece, mirror it and align it to the bottom right of the pink piece. Union all 4 pieces.

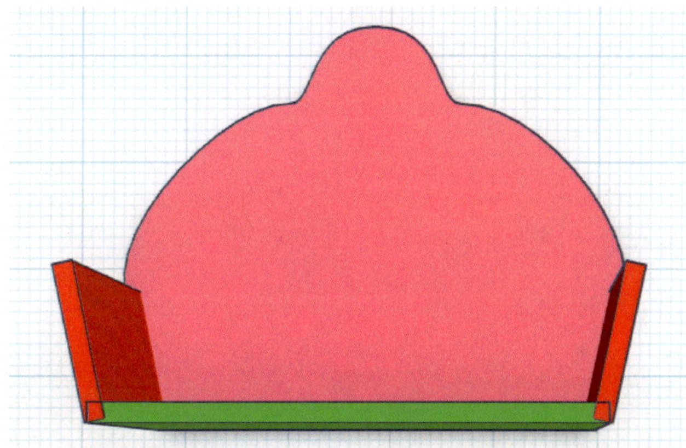

Repeat the legs and touch the back and legs together, centre, then union.

Centre and touch the grate and legs together, then union.

Repeat the grate and legs piece, turn it into a hole and rotate it 180 degrees over on itself.

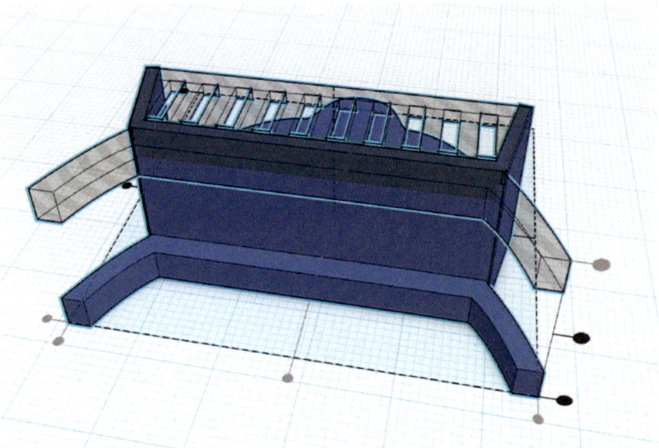

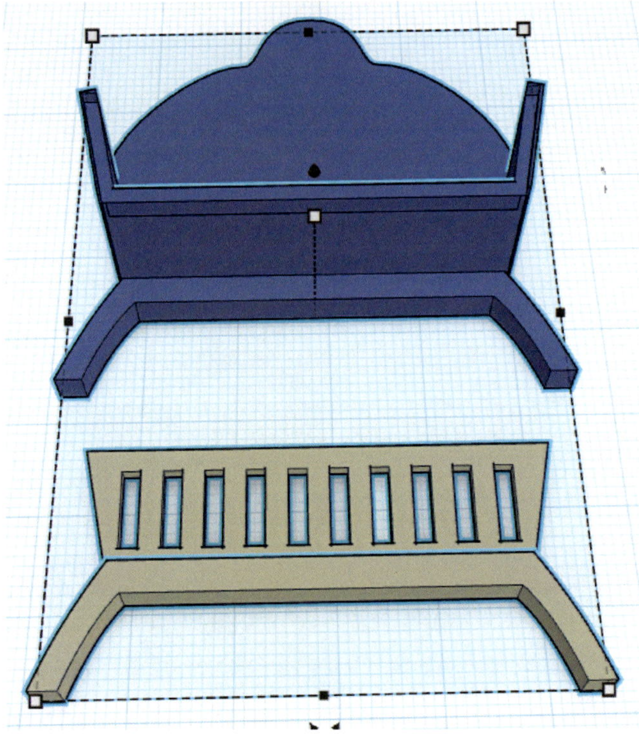

Align it to the vertical top and (notional) bottom of the back piece and union them. This will notch out a small piece making it easier to join the final basket. You now have two pieces ready to export and print.

## Pro tips

*In these projects I give you finished measurements to save frustration but many are arrived at by trial and error. For the smoke chamber and tube I had the rough idea then tweaked everything until they fit perfectly and left a wall of material thick enough to 3D print.*

*If you are designing something new don't worry if it looks fine on screen but you find a weakness somewhere or a piece that doesn't print without supports. Just keep tweaking until it is right.*

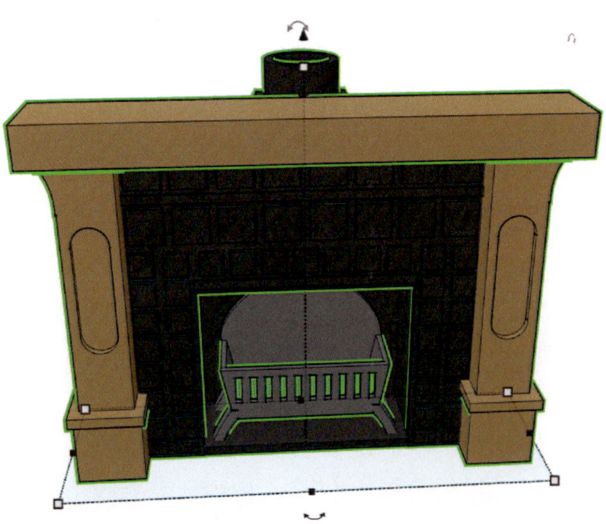

# Bay Window

## This project teaches you how to:

- *Makeshapes and recesses to hold them*
- *Use Tinkercad and Inkscape together*
- *Use Inkscape to map out component shapes*
- *Adding optional decorative elements direct in Tinkercad*
- *First use of the slicer and printing*
- *Start to make decisions on your own designs*

# Bay window

As I often do I started this bay with an accurate sketch. This not only let me design the bay but I used it to transfer proportions and levels to a later project for the doorway and columns. It might seem like a lot of work but it gives you a solid base and measurements for each

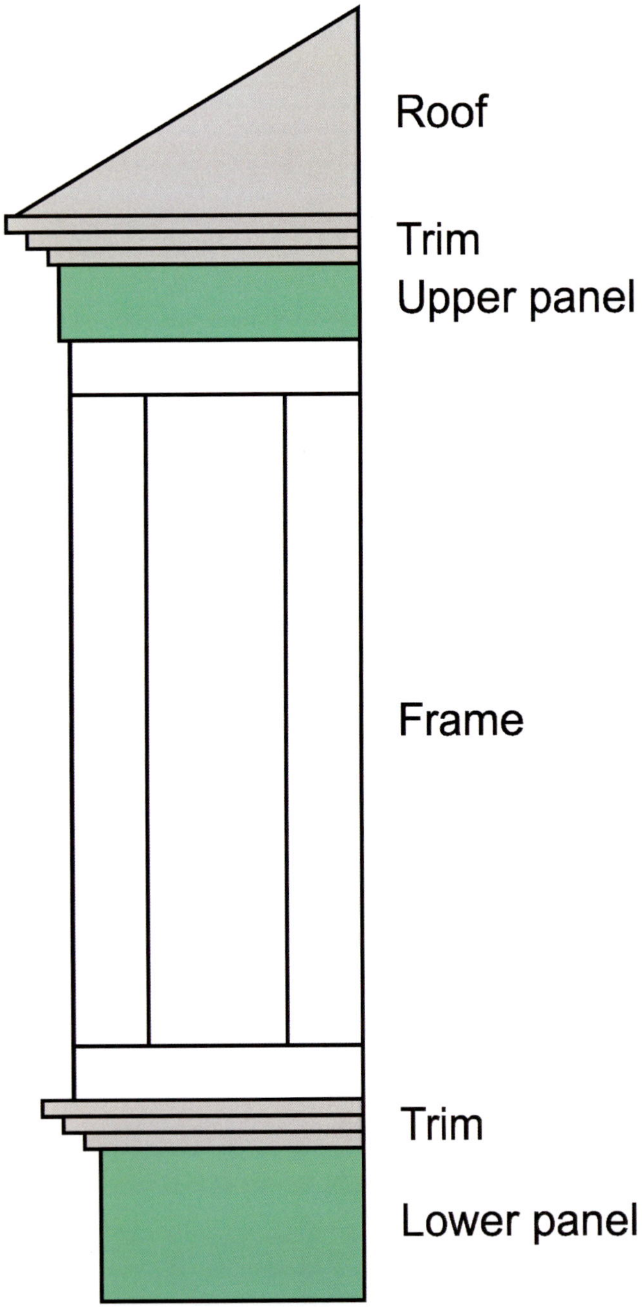


piece you design. I know what size each piece will be both in depth and height. I can now choose a width suitable for the building I am modelling, which is about 210 mm (appx 8.25″) wide by 67 mm (2.5″) deep and totals just under 240 mm (9,5″) high.

Where possible in this book I use the correct names for parts, but in this case the terminology where available is extensive and difficult so I am going to simplify as per the diagram.

I based this window on a bay with 135 degree bends though 90 and 150 degree are popular options too.

I show options for a flat roof, a simple covered sloped roof and a complex tiled sloped roof which is in the photos.

In general I also design each piece with a small exposed section where rain can drip down to the piece below. It is important in the real world so might look strange if not copied here.

I make a basic shape for the frame holding the movable window inserts. I use this shape as the base for all the other parts of the bay, normally in 3mm increments of size.

To make the initial parts it will be easiest to start in Inkscape as it is easier to move the corners out an exact amount there as we scale it. Scaling a shape in Tinkercad would also move the inside measurements which need to keep static. Anyway we'll show that in detail as we go.

Let's start with a box of 11 mm wide by 74 mm and a box of 105 mm wide by 11 mm.

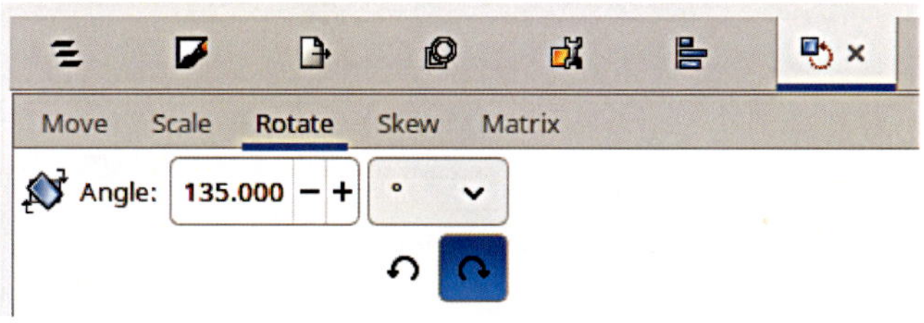


Rotate the taller box by 135 degrees, and move it's top left corner to the top left corner of the wider box.

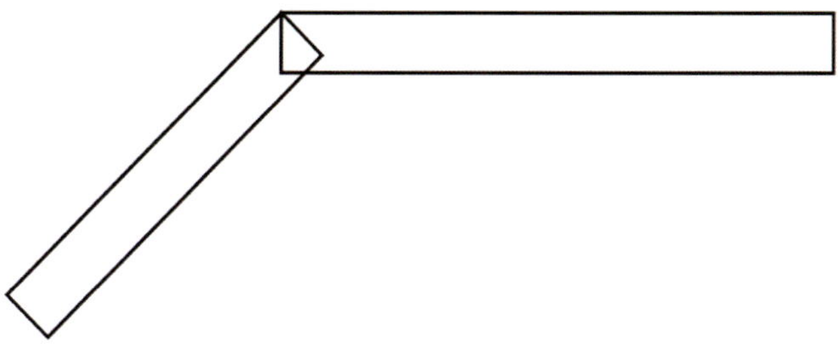

Copy the rotated box and click the mirror button on the top toolbar. Move this to the top right corner of the wide box.

Now draw a box underneath the three and touch the bottom left and bottom right corners of the rotated boxes.

Select and union the three top boxes Path -> Union.

Select the resulting shape and the box underneath and Path -> Difference to cut the shape flush to the bottom.

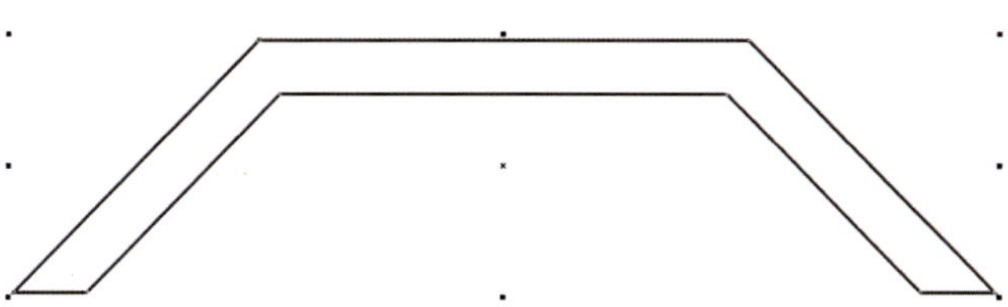

We now have our base shape for the bay components. Using this guarantees alignment and joins all the way up the structure.

You can export it at this point and label it "bay basic".

Next we will make a copy and "chisel" the grooves into it for the window glass to move.

Make a copy of bay basic. Make a box of 90 mm wide by 6.5mm high. Align it centre horizontally to the copy and use the arrow keys to get it near the top.

Draw a small box 28 mm wide by 11 mm, colour it green from the swatch at the bottom, and align it top, centre to the copy. This can now be used to align the 6.5 mm box centre vertically inside the copy.

Draw a box 47 mm by 6.5 mm.

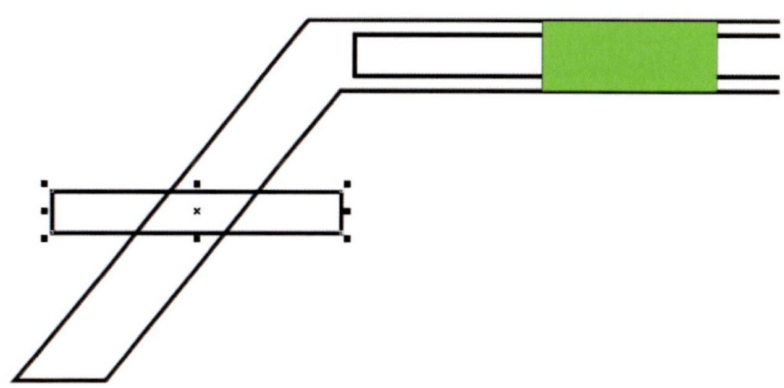

Rotate it to 230 degrees. Duplicate the green box, rotate to 230 degrees and align it vertical and horizontal centre to the 47 mm box.

Move it roughly into position as shown in the next image. It doesn't have to be perfect yet, we are about to do that automatically.

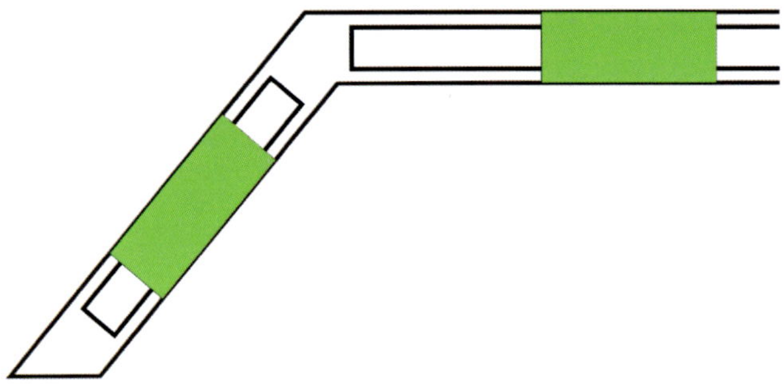

Draw a box of 12mm high and position it so its bottom left corner touches the bottom left corner of basic bay. Then pull the box wider using the centre right double arrow until it touches the side of the basic bay.

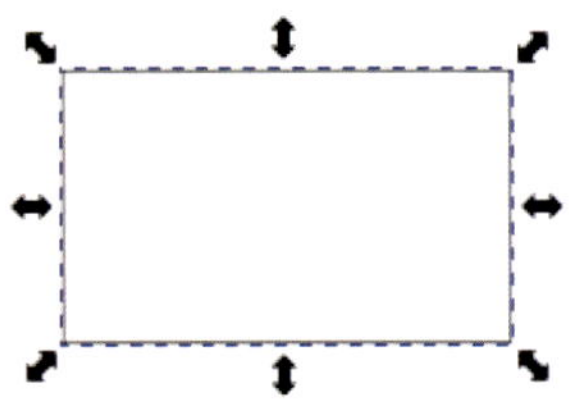

Now select the 47 mm box and green box and drag the bottom corner of the green box to meet the box we just made.

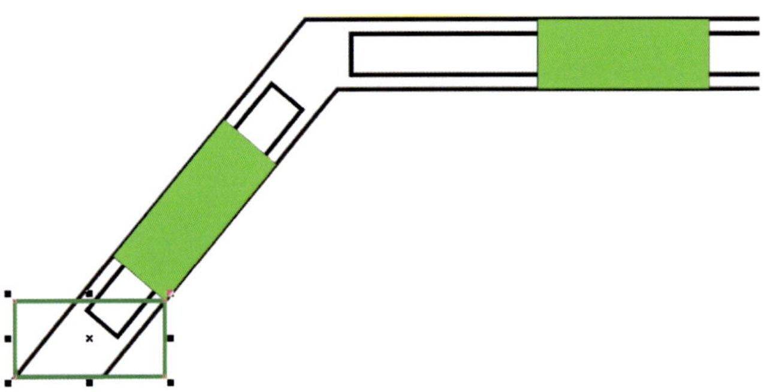

You should now have the 47 mm box perfectly aligned inside bay basic. Select the 3 boxes, copy them and mirror them horizontal. Drag to the bottom right corner to do the same at the other side.

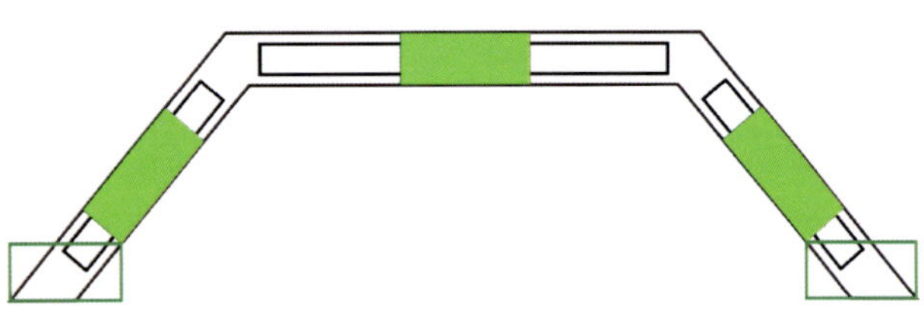

The two 12 mm high boxes at the corners were just guides and can be deleted now. The green boxes will be used again later.

Select the left hand 6.5 mm box and the bay basic shape. Click Path -> Difference. Click yellow from the bottom swatch and the shape should now have a white hole in it. Repeat for the other two 6.5 mm boxes.

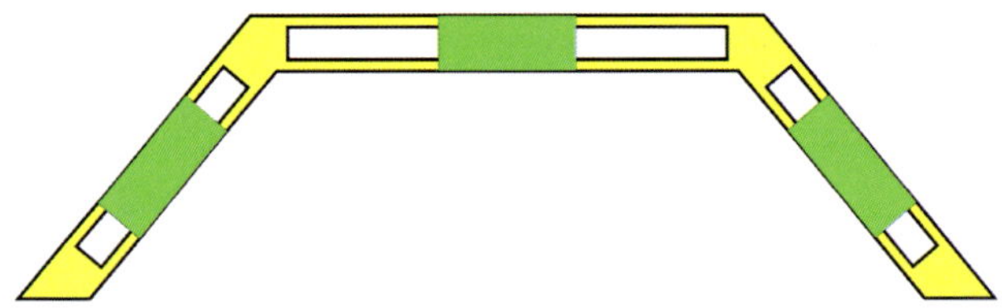

Export the yellow shape as "bay grooves". These will sit top and bottom to hold the glass panels.

At this stage we have a widow sill (bay basic) and these grooves. All we need now are uprights to make the frame. We can adapt the green boxes to cut into basic grooves and just leave 2 uprights at the centre and edges.

Make the green box at the top 84 mm wide by 12 mm.

Remove the rotation of the green boxes at left (-230) and right (-310). Make them each 40 mm wide by 12 mm.

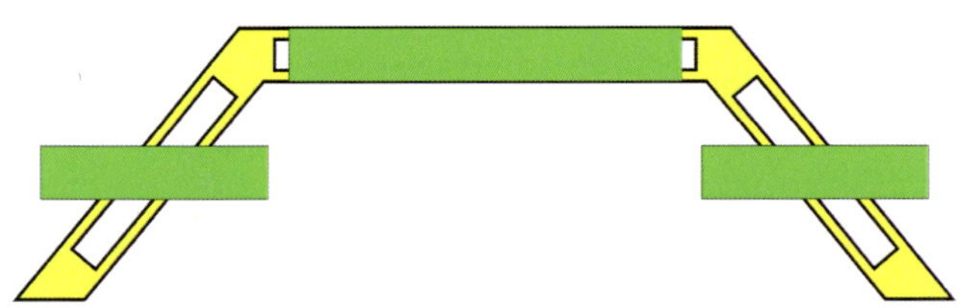

Then re-apply the rotation. They should now sit inside the shape bay groove leaving just a little notch at each side of the glass grooves.

Individually select the green boxes one by one with the bay groove shape and click Path ->

Difference. You should be left with the four uprights shapes we need for assembly. Export them as "bay uprights".

We now have all the parts for a frame that can hold the glass panels. We need to add the panels now. I have chosen to simulate a sash window in two parts for each opening.

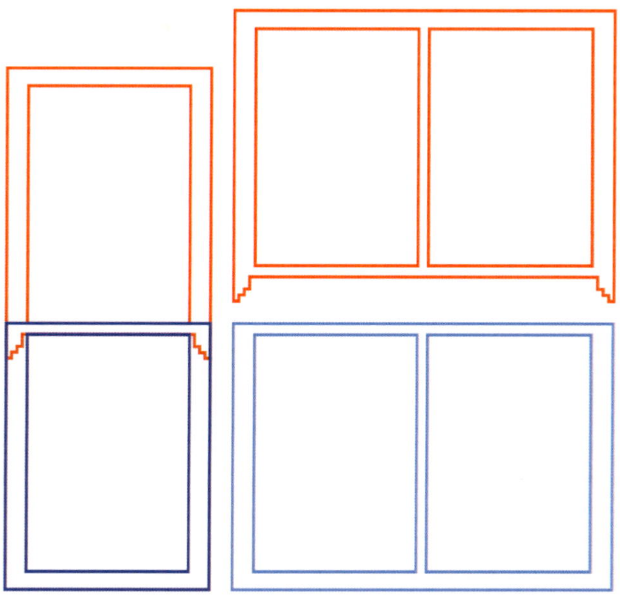

I have sketched the top parts in red and the bottom in blue and shown where they will sit overlapped in the frame.

There is also a little decorative step down at each side on the top pieces.

Looking back at the grooves we cut, we can see the main opening was 90mm, so leave 1 mm leeway and make the panel 89 mm wide there so they can slide into place but will still be held in by the groove in the uprights.

The side groove is 47 mm so make the panels 46 mm wide.

*The forward planning of the initial diagram now helps us. I intended the uprights to be 120 mm high, and the groove and sills to be 10 mm top and bottom. The groove would be 5 mm each, so we have a hole of 130 mm high and should make the panels *total* 129 mm.*

Make the top and sides thicker than they will appear to account for the groove they will sit in all around. The diagram is roughly how they will appear in place more than how they will be constructed.

Let's start at the top, side, window. We can extend this later to create the centre window without having to repeat the detailing work.

Make a box of 46 mm wide by 73.25 mm. Put a box of 32 mm wide by 50 mm inside it, centre it horizontally and vertically. Put another box, 34.25 mm wide by 5.25 mm aligned bottom, centre, of it.

Make a box 2.5 mm square and drag it to the left hand bottom corner of the lower box. Draw a box 1.25 mm square and drag it to the top left corner of the 2.5 mm box.

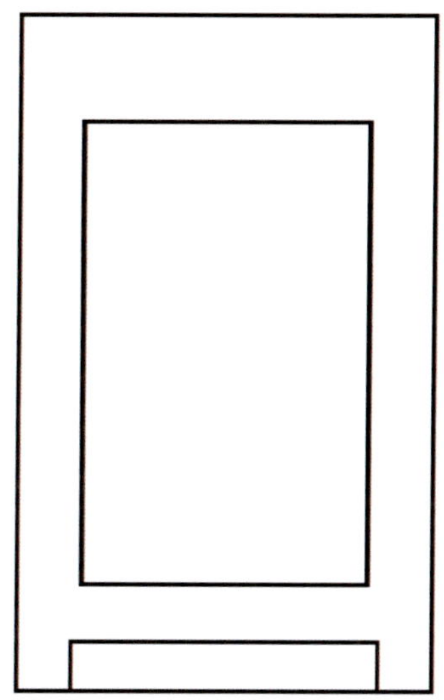

Select them both and click Path -> Difference.

Duplicate this shape and mirror it horizontally to the other side of the lower box.

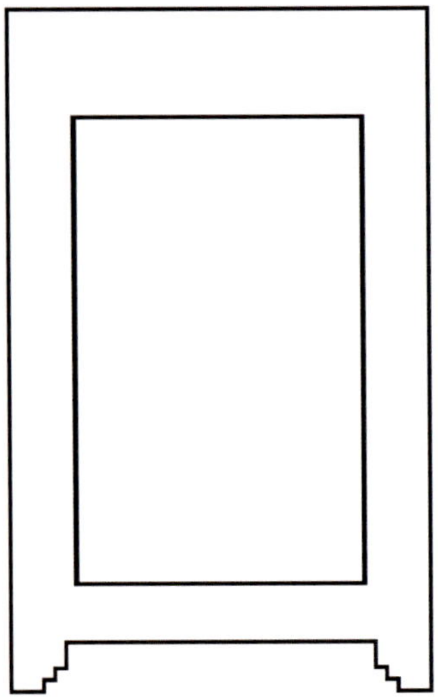

One by one select the shapes and inner boxes with the Large box and click Path -> Difference to cut them out.

We now have the finished upper side panel, and can use it to make the larger centre panel.

Duplicate the shape.

(It might be easiest to draw a box of 89 mm to and move its left edge to the left of the shape to use as a guide first.)

Use the node tool. Click on the shape and select all the right hand points. You can use the right arrow key to increase the size until you reach 89 mm wide.

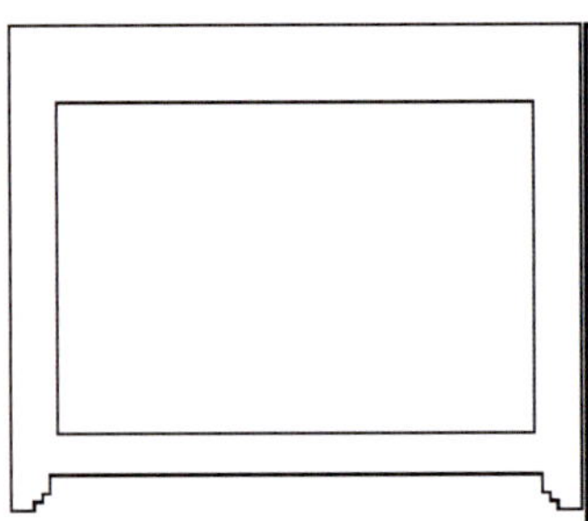

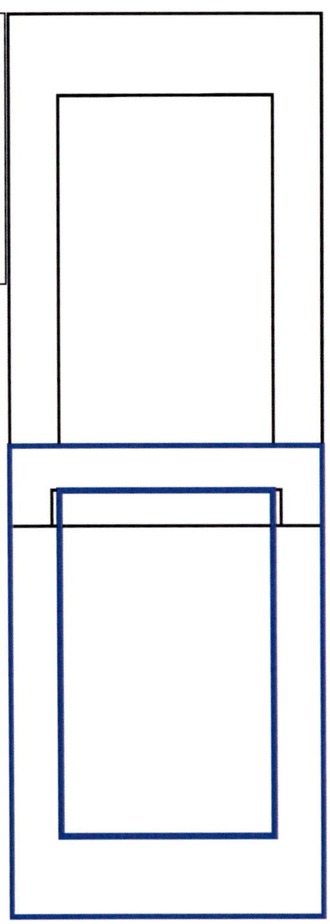

Now for the bottom panel. Draw a box 46 mm wide by 67.25 mm, and a box of 32 mm wide by 49.5 mm inside it. Align the centre horizontally.

Drag the boxes up to align with the end of the glass in the upper frame. Drag the inner box to the bottom bar of the top panel.

Check they are still aligned centre horizontally then select both and Path -> Difference.

Repeat the process of copy and enlarge as we just did at the top to make the centre bottom panel.

To divide the centre glass draw a box of 2.5 mm wide by 67 mm, align it centre horizontally and click Path -> Union.

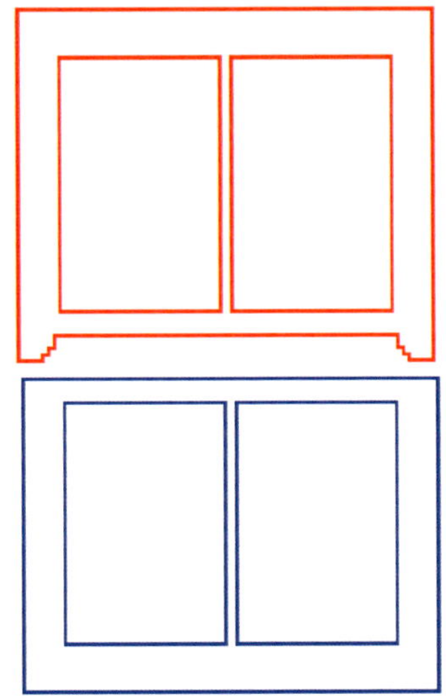

Now we can start assembly in Tinkercad.

Import the shape "bay basic" and make it 5 mm high.

Import the shape "bay grooves" and make it 5 mm high.

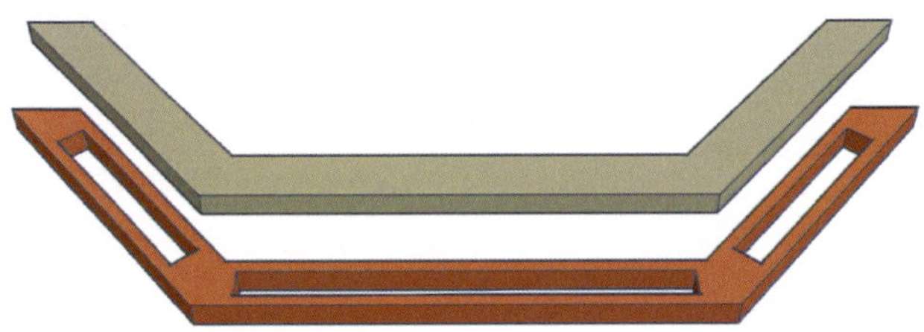

Import the shape "bay uprights" and make it 120 mm high.

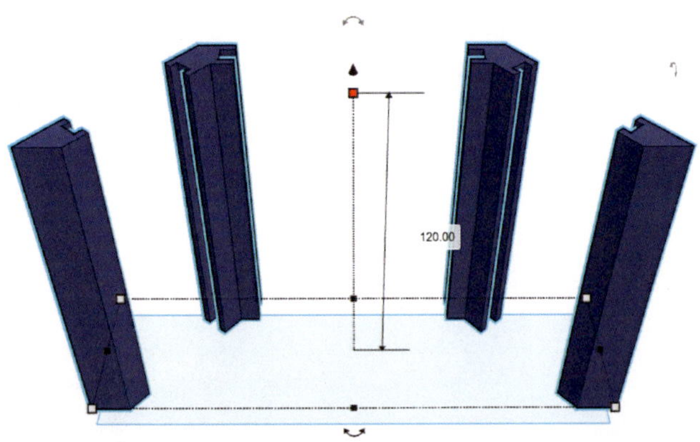

Raise "bay grooves" to 5mm by clicking on it and dragging the cone at the top.

Align it centre to "bay basic", be careful not to change the vertical axis.

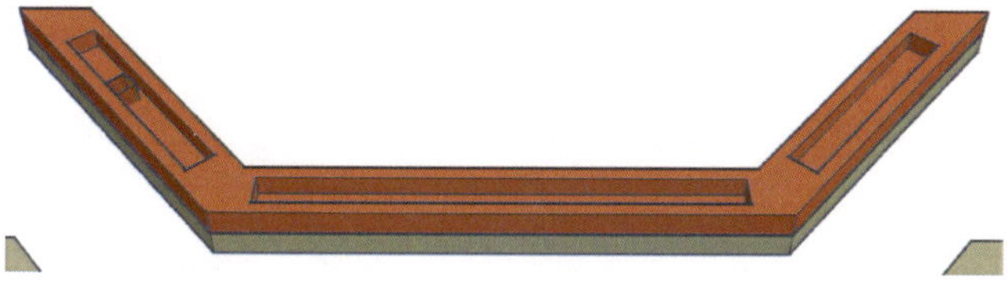

Make a copy of the two shapes and mirror them horizontally. This way even if you made an alignment error along the way, it will still fit at the top.

Raise "bay uprights" to 10 mm, and align centre to "bay basic" and "bay grooves".

Import the window glass panels and make them 2.5 mm high.

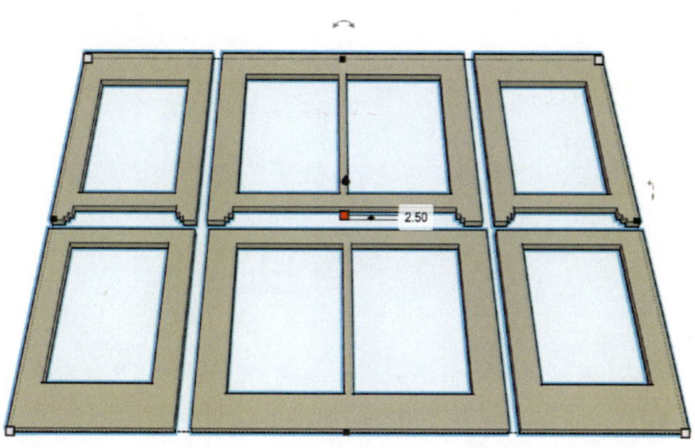


## Slicing

The parts are ready for slicing and printing now. Slicing is explained in the appendix, but for now you don't need to fully understand what it is and how it works just as long as you do it. Depending how big your bed is you may be able to print all at once or have to go in 2 batches.

Export each part from Tinkercad and bring them into your slicer. I will use the free Prusaslicer for the example.

Open Prusa slicer and press the Add button on the top bar to add your STL files to the print bed.

When you have added the combined uprights and groove pieces, click the Arrange Current Bed button to automatically move them away from each other for printing.

Check the options for your printer at the top right. There are often options for speed and quality here too plus type of filament. Note I went for 20% infill so the frame has some strength but isn't solid filament.

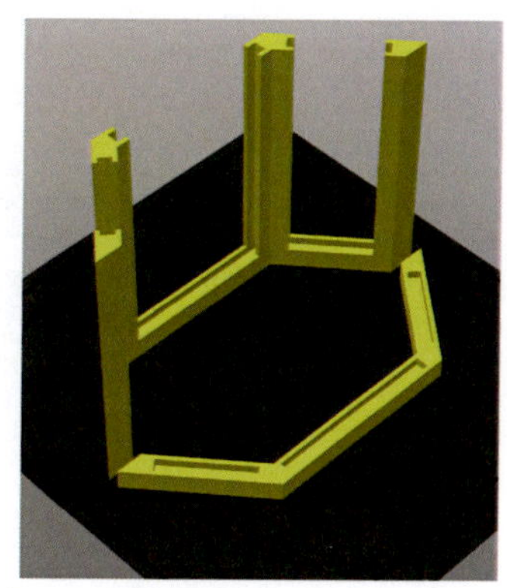

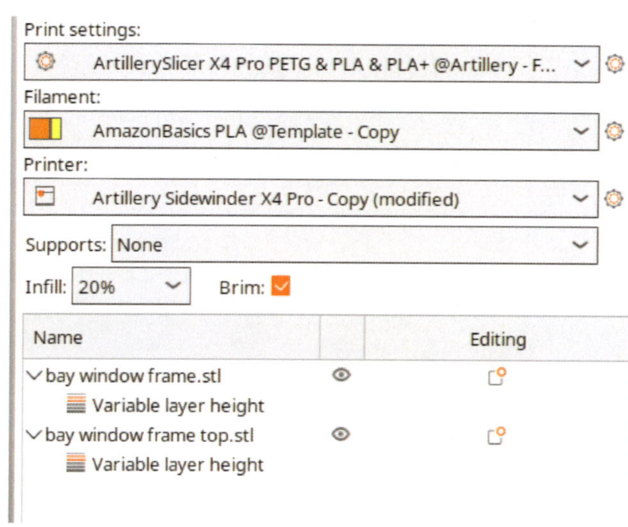

I also ticked Brim which adds a small extra width to the first layer to help stick to the bed, but breaks off easily after printing. The final option you can see I added is Variable Height available from this button which will look for fine detail in the layers and adapt to give the best print possible at those parts but speed up elsewhere.

We can now press the Slice Now button at bottom right.

If all went well you should see a screen with info on how long it will take and how much filament will be used.

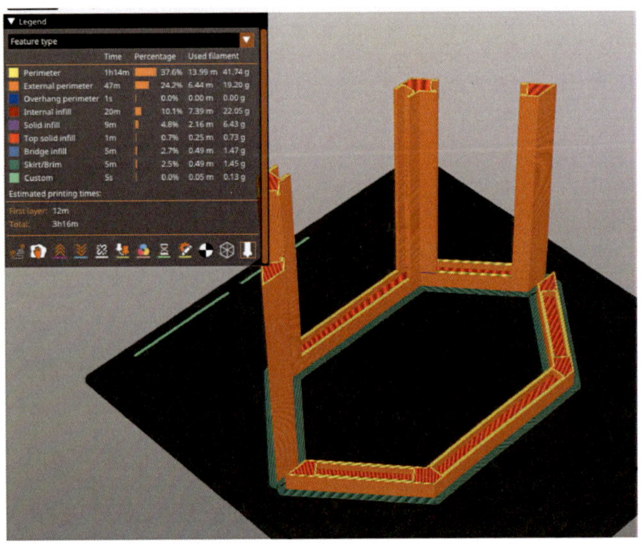

If you get any warnings about empty layers etc. do not proceed. Go back to Tinkercad and ensure everything touches and is aligned with no mid-air gaps.

OK, now we press Export G-code and it is ready to go. Send the file to your printer over the network or via USB stick and print.

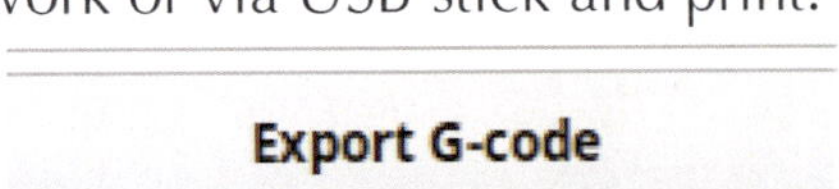


## Decorative trim and panels

We have the tricky parts of the bay in the bag. Now we just need to bulk it out and add trim and panels.

Go back into Inkscape and open a copy of "bay basic". We will now modify it to make all the other shapes we need, including the flat roof option.

I intended all the pieces to step in or out by 3 mm so the easiest way to achieve this is to go to Edit -> Preferences then to Behaviour -> Steps and set "Arrow keys move by" to "3" and "mm". Click the top right X to leave.

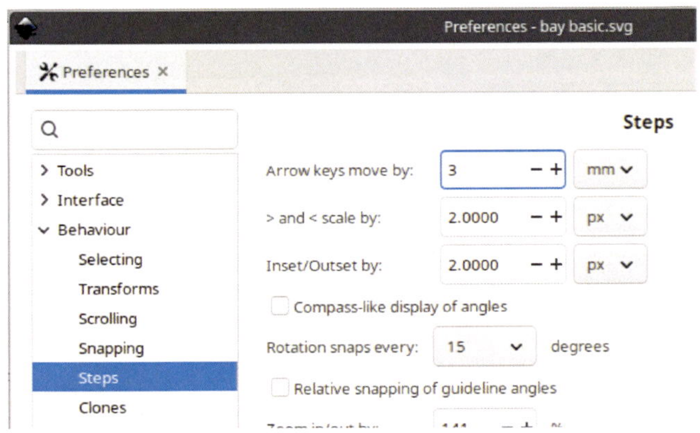


Colour bay basic black, then make a copy/paste and colour it red. Click the node tool. At the bottom left node click on it then press the left arrow, at bottom right press the right arrow. Select the two upper nodes and press the up arrow.

Export this selected piece as "bay red".

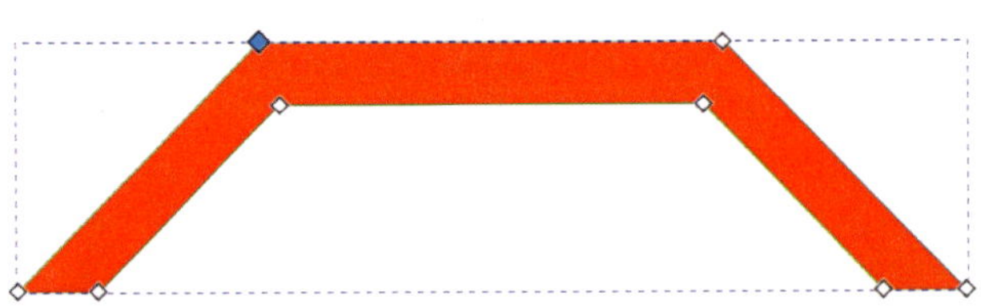

Click Object -> Lower To Bottom then align the centre horizontally and vertical bottom.

Make a copy of the red bay and colour it green. Repeat the nodes and arrow keys, and

Lower To Bottom and align. Export as "bay green".

Repeat for a blue copy. Export as "bay blue". Then a yellow copy. Export as "bay yellow".

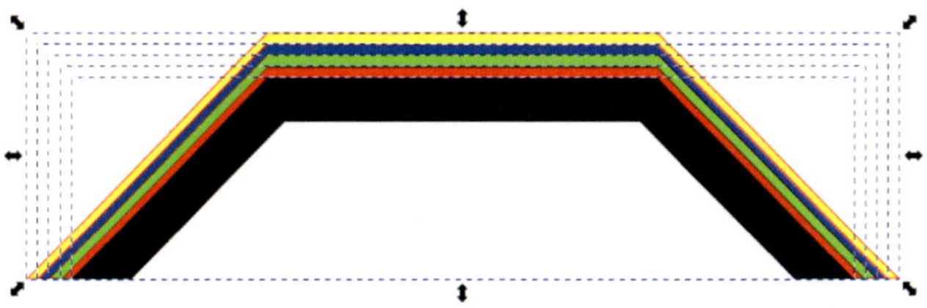

We now have four pieces that are larger than bay basic. We also need two pieces that are smaller.

Make a copy of bay basic and colour it brown. This time with the node tools move the outer nodes in one step with the arrow keys: left side press right arrow, top press down, right side press left. Export as "bay brown". Repeat with dark green. Export as "bay dark green".

This is all we need from Inkscape for this project and the next stage is to build the pieces in Tinkercad.

## Assembly in Tinkercad

*I have recoloured my initial sketch to match the pieces we created in Inkscape.*

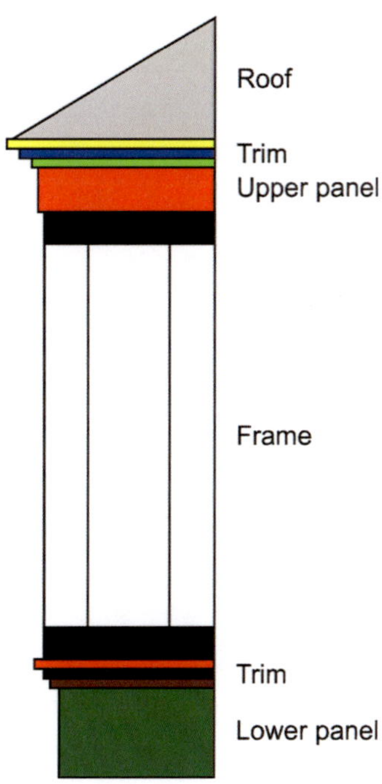


Starting at the bottom we need to import the smallest piece dark green and make it 28 mm high. This will be the lower panel. Now bring in brown, black and red and make them each 3 mm high. These will be the trim. It might be easiest to colour each piece as you bring them in to keep track.

For ease of printing we will stack them with the largest footprint first, so red will be at the base. Raise dark green by 9 mm off baseline. Raise brown by 6 mm. Raise black by 3 mm. Align them all together to the centre and back left. If you like the simple, clean design you can union them at this point and export to STL. If you want to add beading leave them for now and move on to the upper panel and trim.

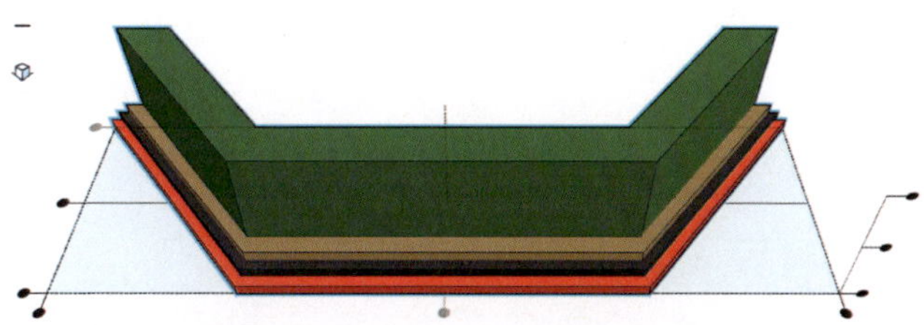

Import yellow, blue and green and make them 3 mm high each. Make blue 6 mm high, green 9 mm high. Import red and raise it to 9 mm, then make it 14 mm high.

Again at this point align, and if you want a simple panel you can union and export to STL now, or for more ecoration add the simple beading to the faces of the panels.

## Beaded panels - optional trim

Starting with the centre panel we are going to make a beaded trim of 76 mm wide by 22 mm.

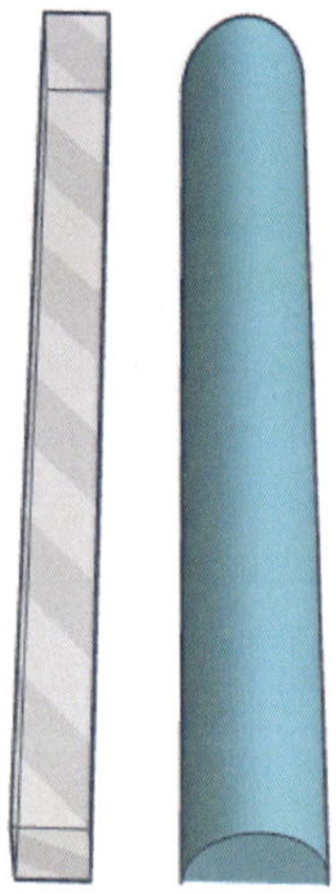

To make the trim we take the semi cylinder and size it 3 mm wide by 22 mm and 3 mm high.

Add a hole cube and make it 1.5 mm wide by 22 mm and 3 mm high.

Align it to the centre and left of the semi cylinder. Union them.

Now make a cube of 76 mm wide by 22 mm and 1 mm high.

Align the trim we created to the left edge.

Repeat the trim and mirror it horizontally. Align it to the right of the cube.

Duplicate the first trim, rotate it 90 degrees and make it 76 mm wide. Align it to the top of the red cube.

Repeat it. Mirror it vertically and align to the bottom. We should now have a trim on all sides that is seamless.

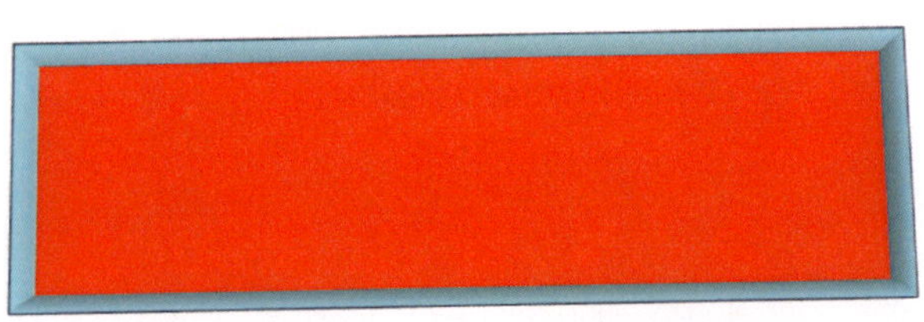

Select all the pieces and union them.

Rotate them 90 degrees to stand up on the bottom section.

Align this piece to the dark green piece, vertical centre, horizontal centre, and to the front.

Looking from the top move it out from the green piece by 3 mm, using the arrow key, until it perfectly touches.

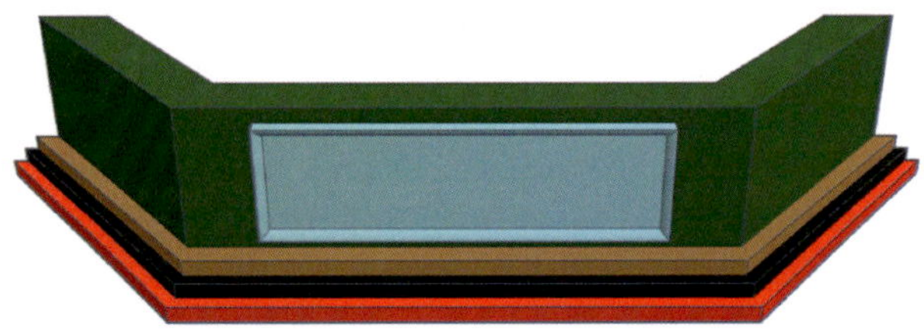

Make a copy of the trim piece and move it away from the current one, we are going to shrink it evenly to fit the side. Unbundle, then click off and click on to the red cube. Make it 37 mm wide by 22 mm. Make the top and bottom pieces 37 mm wide. Align the wayward side piece back onto the red cube.

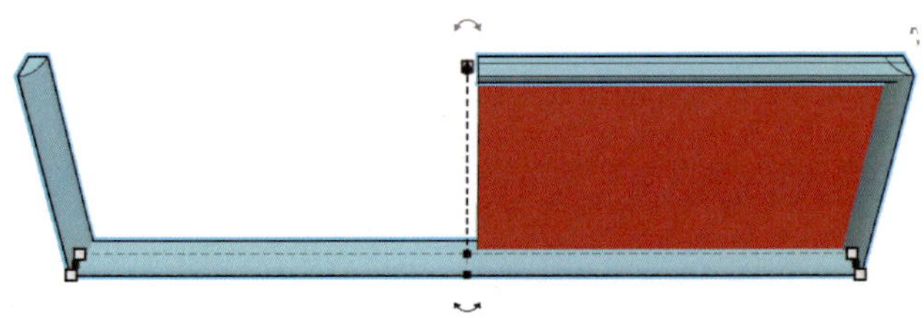

Select all the pieces and union them.

Drag the side panel over to near the side and rotate it 45 degrees. That should be perfect. If it is not then at some stage you have slightly moved a node out of place. You can re-align it by eye or go back and correct the piece.

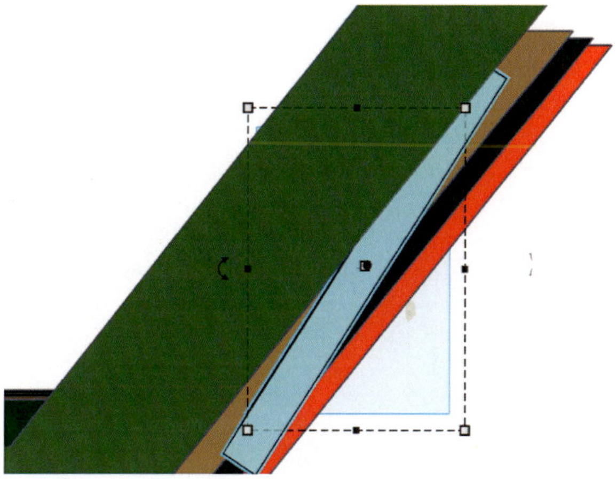

Align the panel to side, front-to-back and vertical centres on the dark green piece. Now move the arrow keys to get it exactly in position. You may need to change Snap Grid down to 0.1 mm again for this as you get close.

Once you are happy, mirror it to the other size, align and position again. Now you can union the whole stack and export to STL for for printing as "Lower Panel".

Now you can Unbundle it and copy all 3 pieces of trim and move them near to the upper panel. We will edit them down to 11

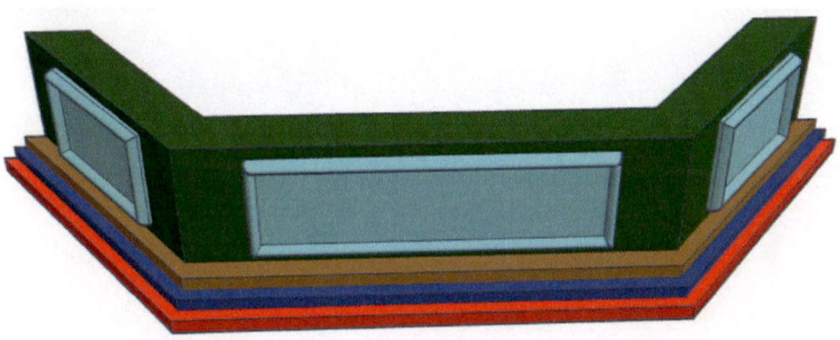

mm high for the upper panel. Unbundle each decorative piece, shrink the height to 11 mm for the red cube and side pieces and re-align everything. You can now align them to the correct places on the upper panel in the same way we just did for the upper. Union them and export to STL as "Upper Panel".

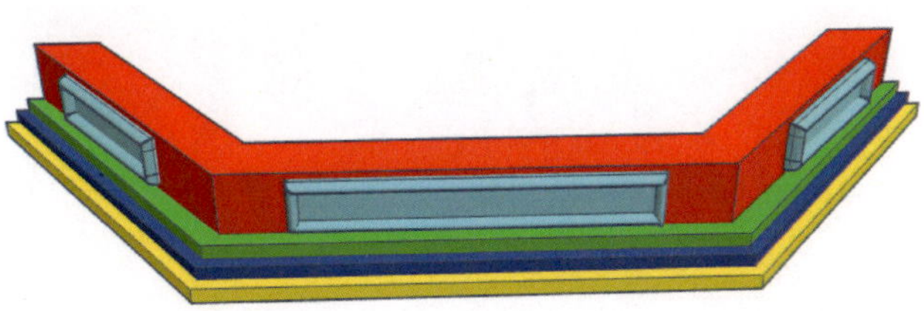

***Note:*** *the beading may need supports enabling in the slicer later to print properly as there is an overhang. I normally use organic or tree supports as they detach easily. These are generated by the slicer to hold any pieces that are printing in mid air.*

## The Roof

We are almost finished with the bay and just have to choose a style of roof now. In the diagram I always intended a sloped roof but in a similar house we lived in, the roof was flat.

## Option 1 - simple flat roof.

This is an easy option, simply copy the top layer of trim back into Inkscape and delete the

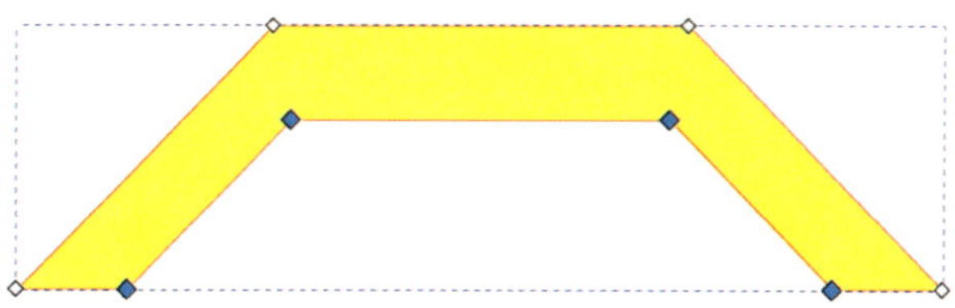

inner nodes to make it fill the whole space. Then we export it as "bay roof", take into

Tinkercad, make it 3 mm high, export again to STL and print it in black.

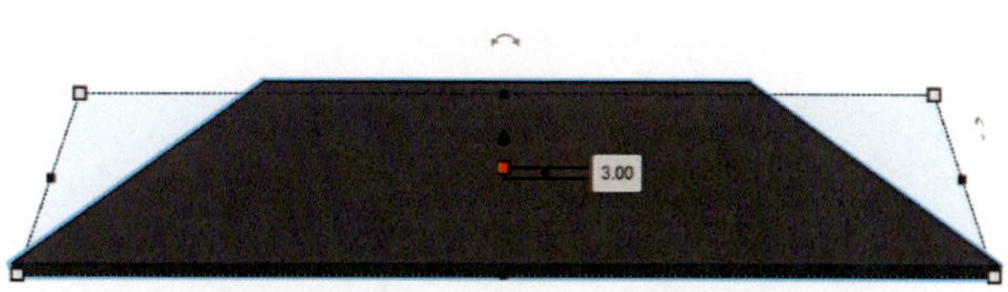

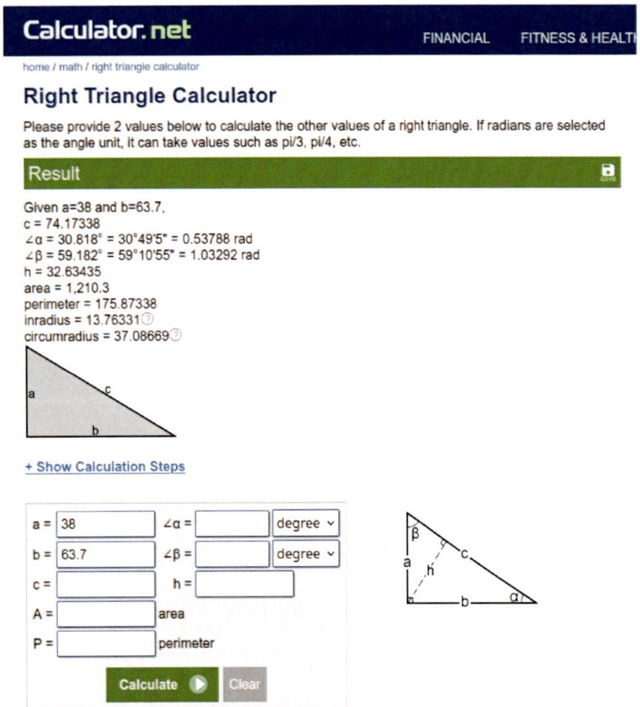


# Optional upgrade: angles

The next step up might be to go around the edges with the semi-cylinder beading, and for that we need to work out some angles. Angles is all mathematics and unpleasant for many so I will do the hard part for you as we need 3 of them. (I am going straight to an online triangle calculator for this). The key angle we need is listed later as corner angle. This will give the angle you need to mitre the edges using hole cubes (more examples are in the first book).

**Height angle.** First of all we need to choose the height of the roof and it's depth. I have gone for 3 mm less then the top trim as I intend to add a 3 mm bead to it at the front to make it flush.

**Corner angle.** The angle of the object side is 135 or 45 degrees.

**Side beading angle.** At the back edge I also put a bead all along the back and need this to slope down to the back corner.

The first calculation example will be the height angle, the slope from back to front. Given the depth is going to be 63.7 mm and the height I chose is 38 mm we get an angle of 30.818 degrees from the calculator (it also shows 90 degrees minus this angle for rotations from the other end).

Side beading angle is 31.784 degrees from 38 mm height and 61.325 mm wide.

# Option 2: simple sloping roof.

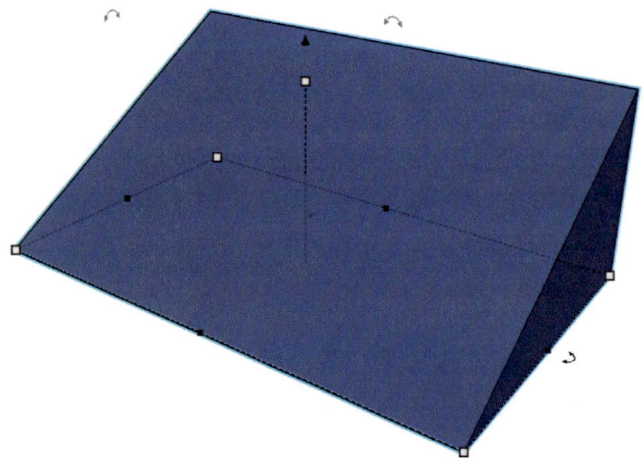

We can start with a wedge primitive, make it 105 mm wide by 63.7 mm and 38 mm high.

This will make the front centre panel.

For the side I ran through loads of possible ways to do it that probably wasted me hours. Here's the simple one I ended up with.

Take a pyramid of 20 x 20 x 20 as it arrives when you place it. Rotate it 45 degrees around the base. Its measurements are now 28.28 mm. Draw a hole cube and make it 28.28 mm wide by 14.14 mm. Copy and rotate it 90 degrees. Align the holes to the pyramid leaving one quarter showing, then Union. Depending which quarter you have left you may need to rotate it later. Make the width 61.235 mm by 63.7 mm deep and the height

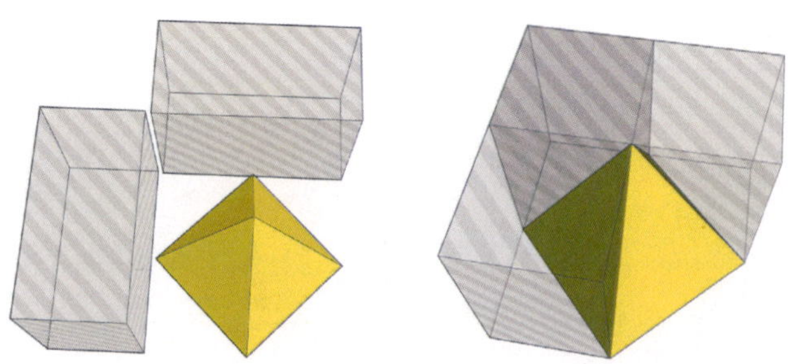

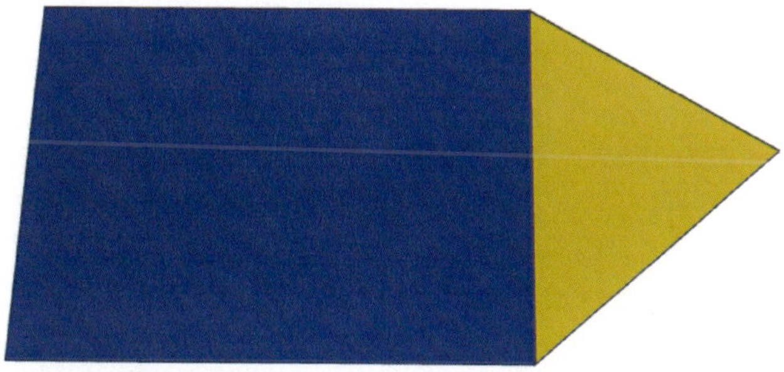

38 mm. Now put it alongside the first wedge and rotate if necessary. OK now mirror the side piece and move to the other end of the wedge.

At this point again we have the choice to union everything, export to STL and just leave it there. Again we also have the option to add detail with beading if you want.

Place a semi cylinder and make it 105 mm wide by 3 mm and 3 mm high. Note you may have to rotate it to go alongside the large wedge.

Carefully place it along the front lower edge of the wedge, then copy it and align it to the top back of the wedge. We will have to use our slope angle now of approx 31 degrees and gently raise or lower it until it fits.

Copy the front piece, rotate it 90 degrees so that it runs front to back and rotate it upright by 31 degrees. You can now align this to front, right of the wedge. Copy it and align to front, left. There are bits sticking up in the air. I don't worry about them at the moment - we will tidy the whole model up in one go before exporting for print.

Make another copy of the front lower bead and rotate it 45 degrees, aligning it to the side. Mirror and repeat for the other side. This just leaves the beading at the back and we have the angle for that (Side beading angle) approx 32 degrees off vertical. Mirror and repeat for the other side.

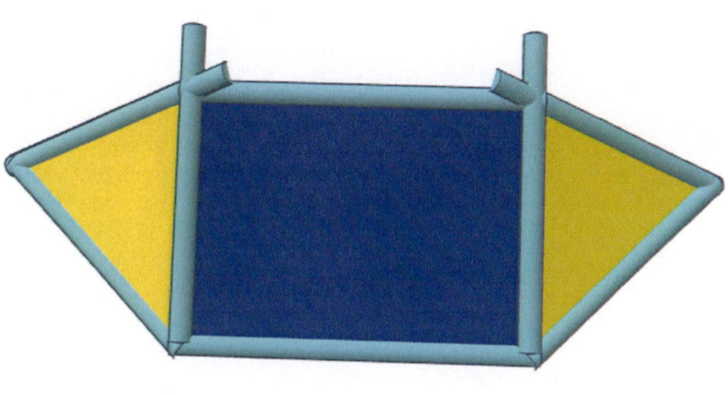

Now for tidying. I made a large hole cube right up against the back edge and copy it below the bottom surface of the model. Finally I copied it again and angled it for the centre slope of approx 31 off vertical and positioned it to catch any stray pieces that go above the plane of the model. Union the whole and you should have a finished roof piece.

Click Fit To All and have a good look around for any mismatched or not level pieces. If you are happy export to STL.

## Option 3 tiled sloping roof.

First we make up all the pieces of option 2 but not the hole cubes to tidy up. We are going to make tile patterned pieces to put onto the faces of the slopes.

You can do this in Inkscape or Tinkercad as it is simply a matter of making a tiled section 105 mm wide by 74.2 mm (thanks Pythagorus). Inside this we will fit 4 and a half tiles wide by 4. Leaving a gap of 1 mm or so on all sides. This makes a tile approximately 25.4 mm wide by 18 mm. Simply arrange and align in the software of your choice.

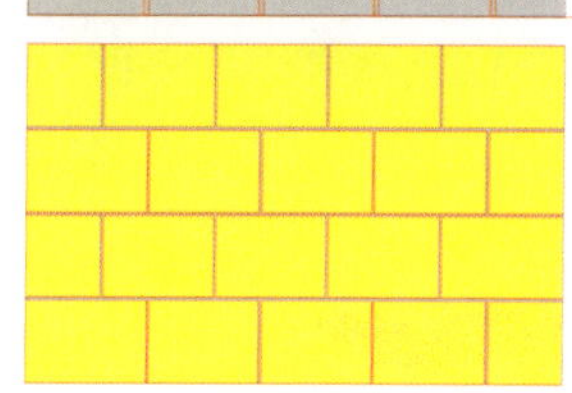

Make the base plate 1 mm high and the tiled zone 2 mm to have a subtle effect. Before we place it take a copy for the sides.

Raise it 31 degrees from the vertical and align it to the main wedge piece. You can use the arrow keys to merge it into the wedge and the beading.

OK now for the side pieces. Raise it by 31 degrees from the vertical. This isn't perfect but averages between the two angles we calculated for slope and side and we are going halfway between the two. Rotate 45 degrees around the base and move it into the side. Align it to the bottom of the side piece. Carefully position with the arrow keys until it merges into the beading pieces but still shows the pattern.

At this point it looks like an awful mess with bits jutting out everywhere, but if it looks right at the side beading it is good.

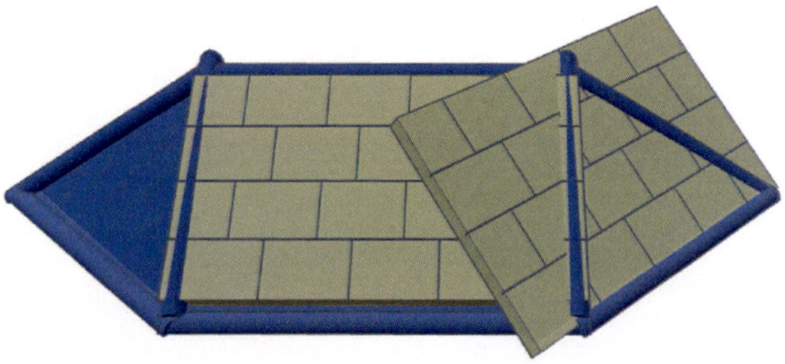

Mirror it to the other side.

Now just like we did in option 2 we can tidy the mess up with hole cubes until it looks perfect. Add them at the back, underneath and at the main wedge slope. Union them with the side pieces and it should look good again.

## Option 4: double height

You can leave the roof part for the first story too and instead build up a double height bay into the upper floor.

## Glass

I normally use a piece of acetate stuck to the window panels for glass. If you want to be able to look from the inside too you can sandwich it bewteen the glass panel pieces and print two copies at 1.25 mm instead of one at 2.5 mm.

***Note:*** *with the two sloped designs you may want to hollow them out inside (just use a copy of the wedges reduced by 3 mm all around and convert to a hole. You can also print it upright on it's back piece and turn Brim on in the slicer to get a fine trace around the outside that you can leave attached and it looks like leading or bitumen paint.*

# Doorway

## This project teaches you how to:

- *Design components that look right together*
- *Make decorative glass from acetate*
- *Make architrave designs simply*
- *Save filament on larger shapes*
- *Basic 2 colour printing*
- *Interlocking shapes*

*At this point I assume that by now you don't need all the handholding and positions of the key buttons in the first projects. If you lose track then the controls for the software are all in "cheat sheets" at the back of the book.*

# Doorway with columns

I love a diagram when I am planning a project. It makes all the measurements so easy later in the assembly. So here I have taken my door profile in Inkscape and copied all the heights across to make a door. The trims are in the same place, the column is the same height and position as the windowframe. The plinth is the same as under the window. I made a design choice here to have a leaded light window above the door and an arch on top of the columns. You can easily swap these out for your preference, perhaps a simple flat roof or single wedge.

The door also has glass panels where I intend to print a design on acetate and also on the leaded light above. If you want you can also 3D print black lines into the exact places the colours change to give a more realistic effect.

There is a tiled step and entrance porch included in this design, again they are entirely optional if you want a simpler look. I have gone quite grand here but only planned one single height bay and and doorway. You can extend the bay up to the upper floor and make the house double fronted quite easily.

My grand bay and door mean the house will have to be equivalent of 4.5 m / 15 feet wide.

We will make brickwork in the next project to fill in the gap between the door and window. We will make tiles for the doorstep and porch. All of these are optional and can be left plain.

For the door surround we are going to make architrave and some top beading we also use later as dado rail.

Thankfully there is no tricky maths in this section.

## Architrave and dado

Let's make a complex looking piece with simple blocks and ovals.

I want something approx 120 mm wide by 27 mm to go around the door frame and be decorative. After looking at various photos on the internet I chose a shape.

You can do this design either in Inkscape (with the option of perfecting it at the end) or Tinkercad using the top down view (it'll make the shape but maybe not perfectly.

Start with a box of 23.5 mm wide by 27 mm.

At its side add another 10.5x24.5, and one of 36.5x19, one of 64.5x13.5 and finally one of 18x10.

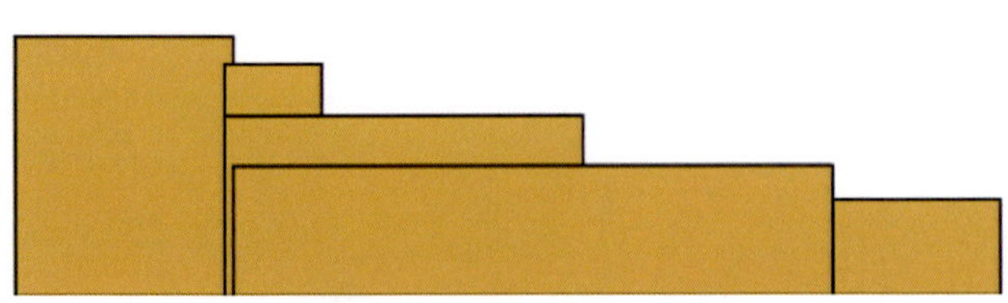

Roughly touch them together as shown in the top diagram and union them.

We will add shapes to this (shown in blue) to make it nearer the shape I want.

Add (by eye) a triangle or rotated box between the first two boxes just to slope it down into the second box but leaving more than it covers.

Add an oval 21.5x23.5.

Add another 3 triangles by eye to slope between the steps.

Finally add an oval 11.5x10.5 to the end. Union it all.

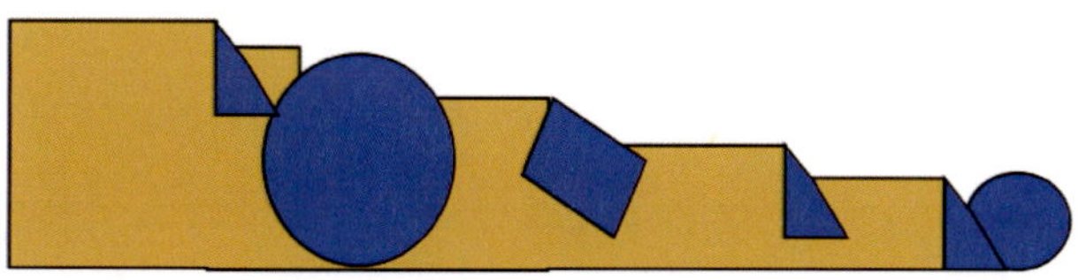

You might just see little pieces that aren't quite right at this point if you zoom in. We don't care yet and we are about to make a couple more.

Now for some cutouts shown in red. An oval of 8.5x10.5 and one of 15x25.5. Path->Difference them (or hole the red pieces in Tinkercad then union).

Finally if there are any rough sections there are two options. At this scale you can just ignore them for a filament print, it really is unlikely to show. If you are a perfectionist you can

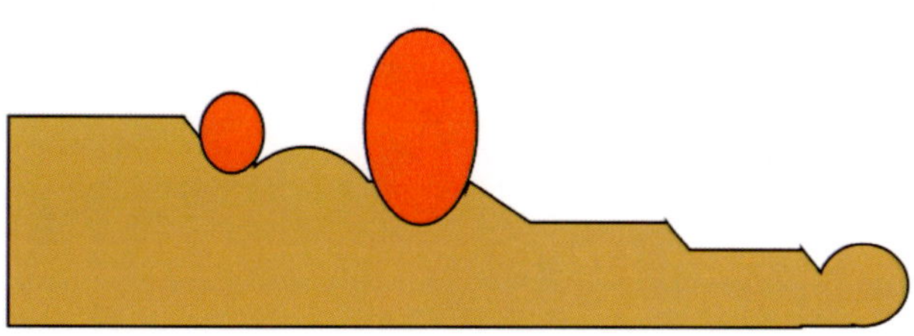

go to the node tool if you did it in Inkscape and twerk or delete nodes to make a pleasing shape.

This is a nice, detailed, piece to run right up the sides of the door frame.

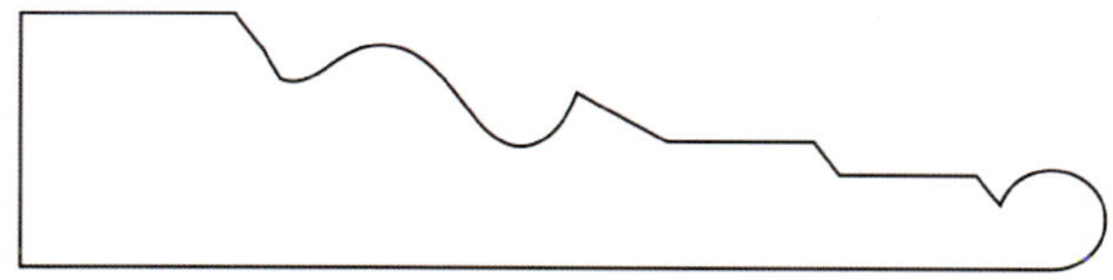

Along the top and around the top window it might be too fussy so we will make a much simpler one which we will adapt later for the dado rail.

Make a box of 10x2, and one of 8.25x1.5 sat on top of it.

Make a circle of 6.35 and align it bottom, left to the first box.

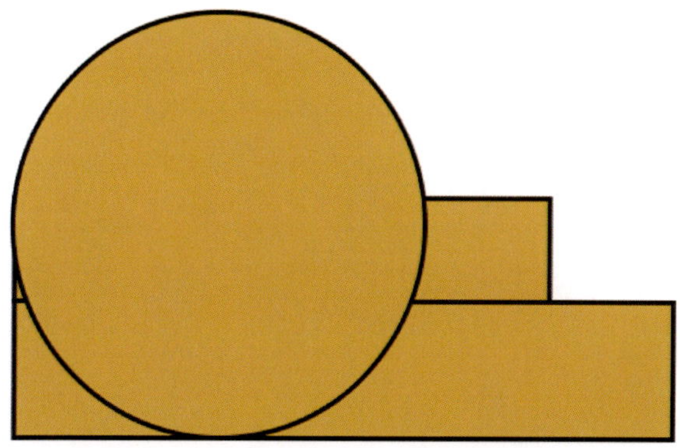

Union the pieces and you can see it is 90% there and certainly useful as-is at 12th scale on filament.

If you want you can tweak it by deleting the inner step nodes and pulling the curve into place. It's a skill worth trying as it allows you to make elegant curves from boxy beginnings.

Export the first shape as architrave and the second as dado and import them into Tinkercad.

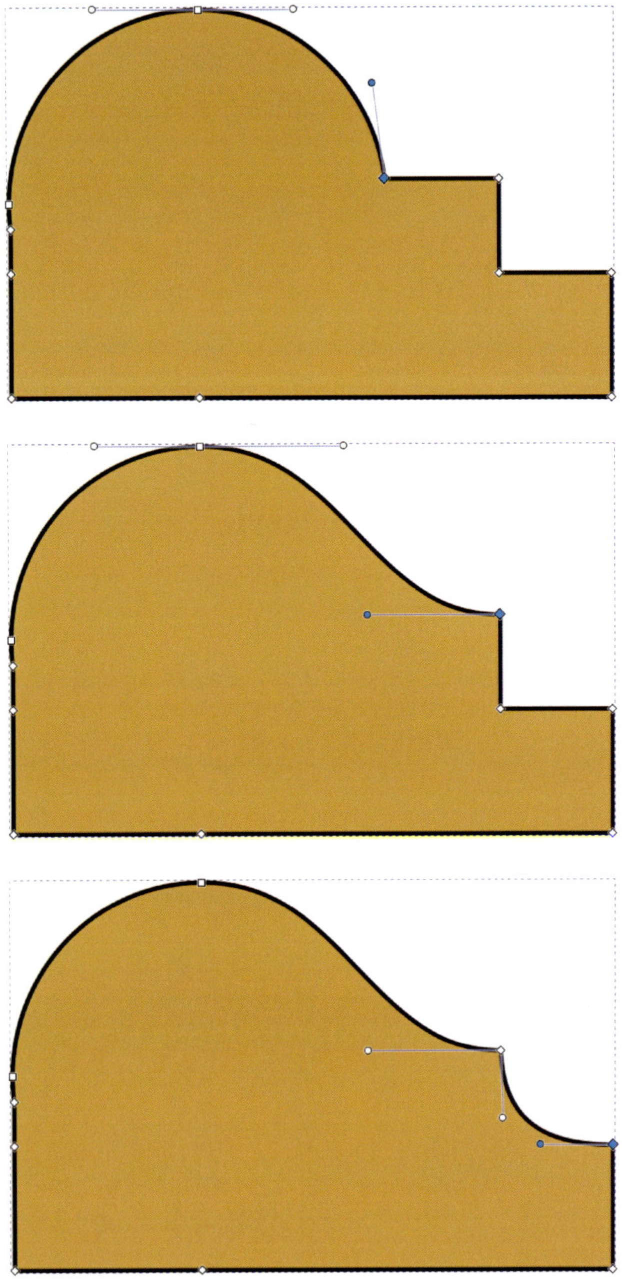

We now have a door frame.

Rather than mitre the joins I have let them blend, it would take near witchcraft levels of skill to make a mitre look good there. A master craftsman could do it in wood but I am taking liberties with the detail visible in 12th scale and just letting them merge.

To make the frame for the window above it, copy the dado piece and move it up and away about 5cm.

Change it to 105 mm wide and centre it back on the top piece on the door frame.

Repeat it, mirror it vertically and move it down near the door frame.

Repeat it, rotate it 90 degrees and align it left to the first sideways dado then top to the window frame.

Repeat this piece, mirror it horizontally and align it left to the door frame dado and then top to the window frame.

Finally align the second dado copy to the bottom of a side piece and you should have a finished frame. Don't print it yet.

## Assembly in Tinkercad

Rotate the architrave 90 degrees flat onto the ground and make it 199 mm long.

Rotate and make the simple architrave 110 mm wide and align it to the top left of the architrave. Repeat the architrave, mirror it and align it top right to the simple version (we'll call it the dado as it will also be used for that later).

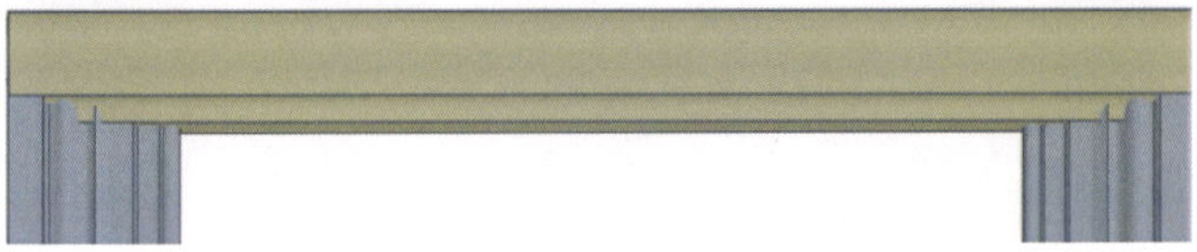

# Door

Now for the actual door with two windows and decorative panels. We can do this one in Tinkercad

The main body of the door is a cube of 86.5 mm wide by 195.75 mm and 3 mm high.

The windows are two hole cubes of 22 mm wide by 93 mm.

Place them inside the body and 18 mm down from the top. Don't worry about exact alignment, we will do that later. We will also add 3mm beading around them later.

Make a hole cube recess 29.5 mm by 8 mm and 1 mm high for a letterbox (which we can print in metallic and add later). Align it to the vertical top and centre of the door but don't union it yet, we will adjust by eye to the other pieces.

We will make a beaded panel 68 mm wide by 54 mm at the bottom.

For this I have used the Round Roof primitive at 3 mm wide and 3 mm high.

The width across the top and bottom is 68 mm. The sides are 52 mm so that they merge into the beading. You can align them all into a box then use the arrow key to move the top up 1 mm and the bottom down. Now bundle them all and move them 7.5 mm off the bottom. Align the bundle to the centre of the door.

Now we have points to work from in aligning the window beading.

The top and bottom piece of beading will be 28 mm wide. Align them centre to the hole cube.

Now use the arrow key to move them 3 mm up and down.

The two sides will be 96 mm long. Align the centre and left, centre and right to the beading and bundle them all with the hole. Repeat the bundle.

You can now align one hole & trim to the left of the bottom panel and one to the right. By eye, move the letterbox up / down to where it feels right.

Select everything and union, check for any errors. If it looks good, export it to "door.stl". Don't print it yet, there are choices to come later.

# Columns

Starting at the bottom the plinth will be a cube 24 mm wide by 28 mm and 20 mm high.

On top of it the first trim piece will be 30 mm wide by 3 mm and 23 mm high.

On top of that a second trim will be 24 mm by 3 mm and 20 mm high.

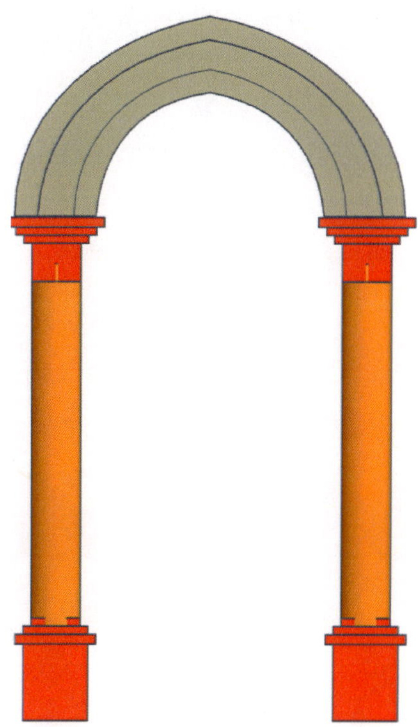

The final trim will be 18 mm x 3mm and 17 mm high. Note all the measurements move in steps of 3 mm. You can align centre and union all four pieces at this point.

In this trim we are going to embed a cylinder hole of 16.4 mm and 5 mm high to hold the column. Rotate it on its side and align it flush at the top, and centre and bottom to the trim.

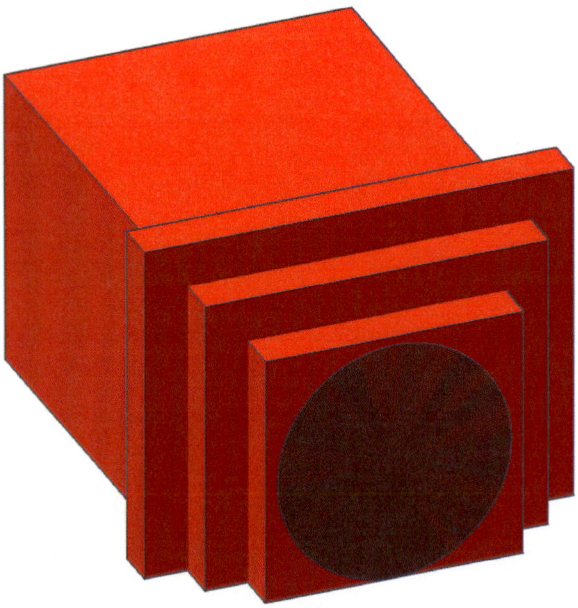

Now make a cylinder of 16 mm and 150 mm high. These are easiest printed upright if separated from the plinth, and with Brim added to make them more stable.

The column top piece is a cube of 18 mm wide by 14 mm and 17 mm high.

The trim is a cube of 24 mm wide by 3 mm and 20 mm high, a cube of 30 mm wide by 3 mm and 23 mm high and a cube of 36 mm wide by 3 mm and 26 mm high.

As at the bottom, recess a hole cylinder of 16.4 mm and 5 mm deep into the part nearest the cylinder. Union the pieces.

You can print the trim separately from the cylinder if you want to print them different colours as I have done, or you can carefully align them as one piece with no need for the hole recesses in the plinth and top.

## Option: simple triangle porch roof

A simple triangle wedge roof is the easiest option.

Simply place one 147 mm wide by 29 mm and 72 mm high.

Make another 135 mm wide by 29 mm and 66 mm high.

Copy it and move it away, make it 23 mm deep instead of 29.

Turn the original smaller triangle into a hole and union it with the large one.

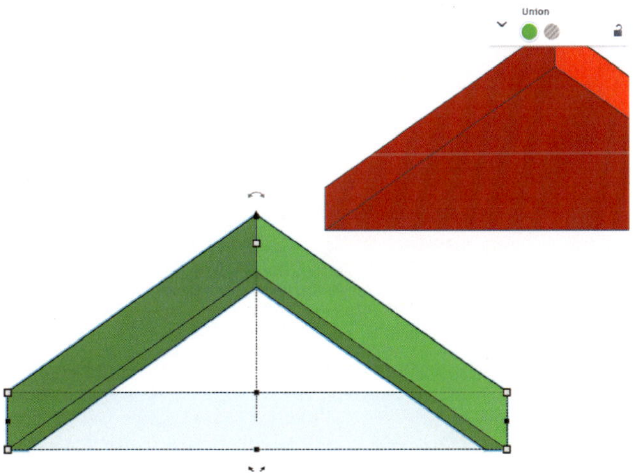


Align the remaining triangles together at the centre and back.

If you wanted to add detail you could cut thin lines into the front of the smaller triangle before union, to look like planks.

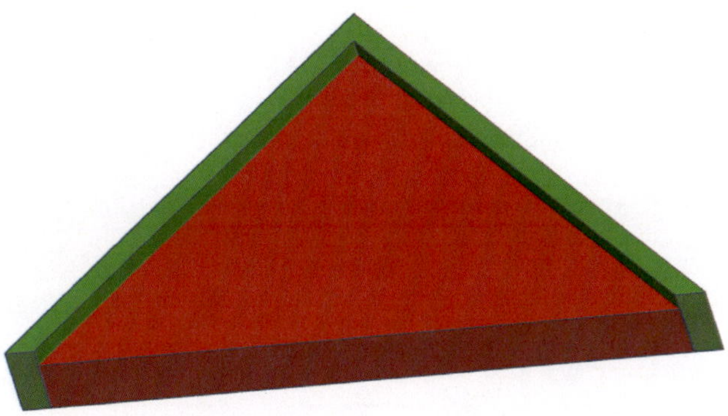

## Option: arched porch roof

If you did the bay project earlier where we set the arrow key distance to 3 mm, set it back no to 1 or 2 mm (Edit -> Preferences -> Behaviour -> Steps).

On top of the columns I chose to put an arch. It may just have been delusions of grandeur but looking at the options of flat, sloping, triangular and arch I decided to do the arch.

On the photos I have seen the arch appears cut into the brickwork and showing part of the leaded light behind it which I liked.

Let's do an almost-Gothic curve.

I chose to draw this in Inkscape as it is relatively simple there to make the shape.

Draw a box of 147 mm wide by 72 mm.

Draw another box .1 mm wide and centre it to the first box.

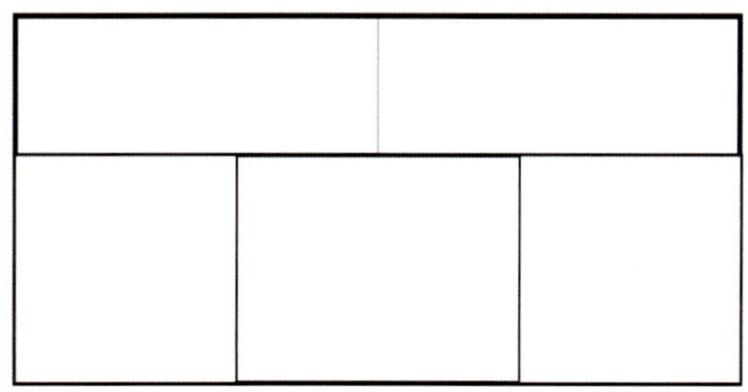

Draw a box 45 mm high aligned to the bottom. This is the bottom of the arch curve.

Draw a box 45 mm wide aligned left and copy it to the right. These are for the curve guide later.

We have the key positions of the arch in place now and can start to make the curve knowing that if we slip there is a marker to push the shape back to.

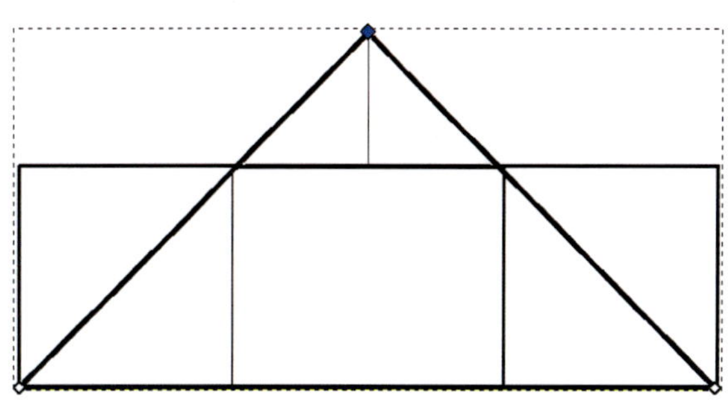

Click the large box then Path -> Object To Path.

Click the node tool, select all the nodes and click add curve handles.

Delete the top left node and move the top right node to the centre line.

Click the top node and click "make selected node symmetric" which will produce a curve.

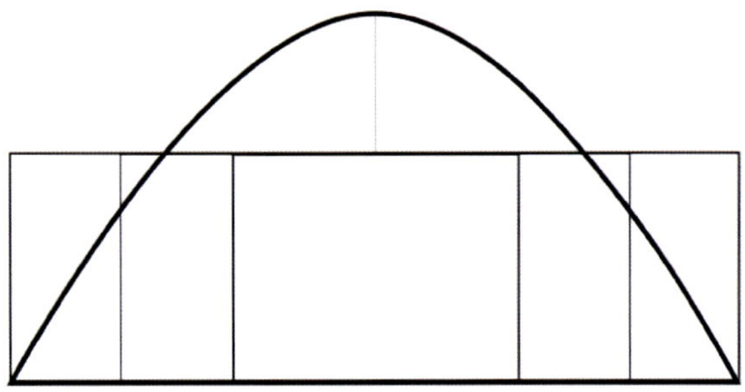

Draw boxes 22.5 mm wide and align them left and right at the bottom. These will give us a guideline for the curve.

Click the arch, click the node tool and pull each side of the arch to the new boxes.

Double click at these intersections to add nodes. Select the two new nodes and press

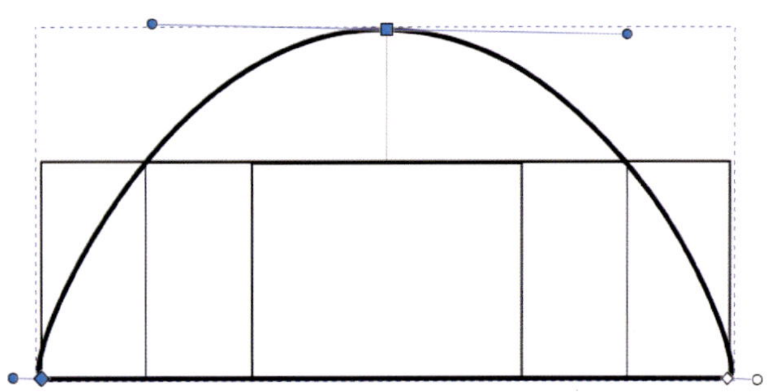

Shift C. They are now cusps meaning they operate independently in each direction. We will now be able to shape the upper part of the curve more towards a point without destroying the lower arch.

To get a point of symmetry between the two sides I made the 45 mm wide side boxes increase to 62.5 mm high and pulled the handles from the top and middle nodes to meet at the intersection of the boxes. You may

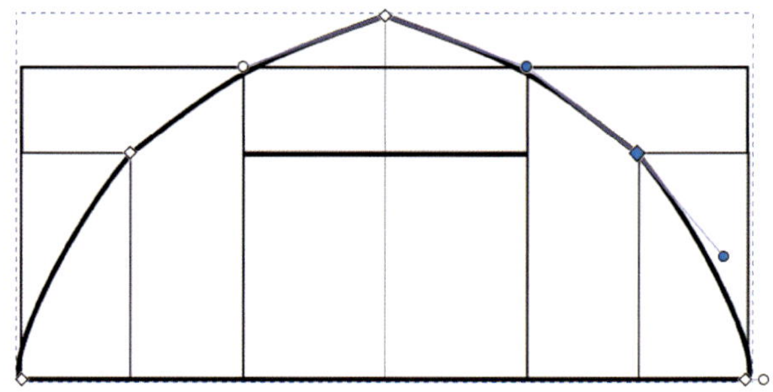

notice the curve doesn't follow exactly through that point, just the control handles.

Now copy the curve and reduce it to 87 mm wide by 64 mm, align it to the centre.

Select the two arches and click Path -> Difference. We now have the base arch, next we will cut a groove into it.

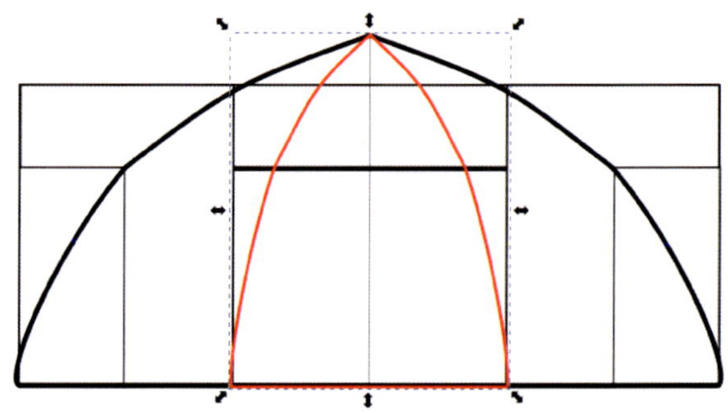

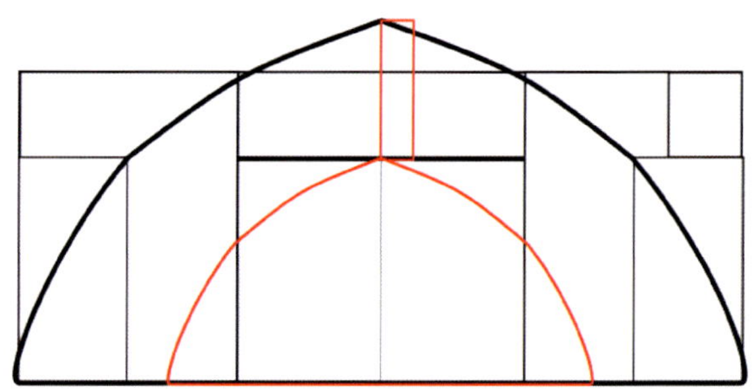

Colour the arch red.

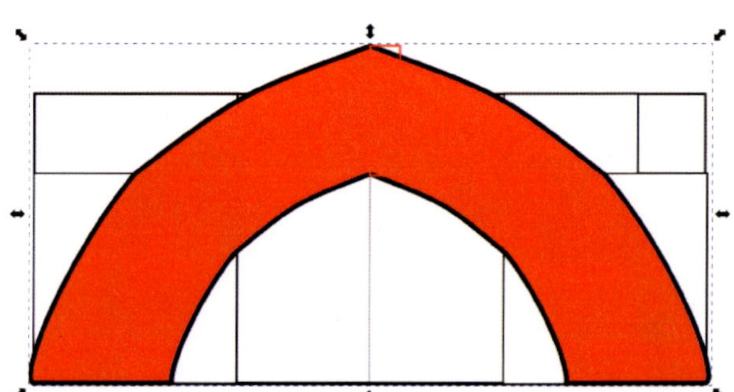

Now copy the curve, colour it green and reduce it to 139 mm wide by 66 mm, align it to the centre and bottom.

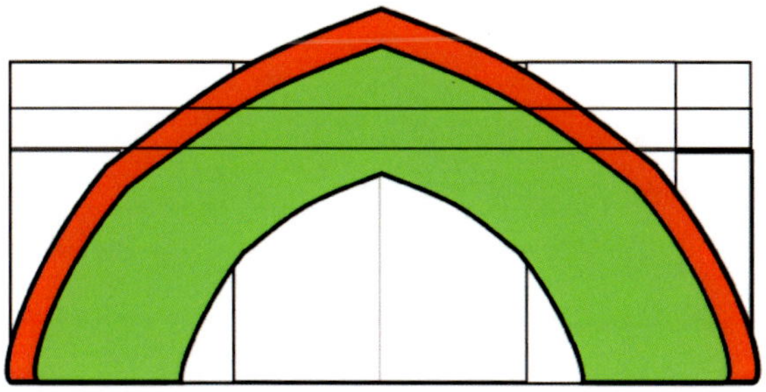

At this point click the inner leg node and move it in 8 mm with the arrow keys, repeat at the other side and at the lower arch peak. There are the two extra nodes and move those around with the arrows until it looks *kinda* right. It doesn't have to be perfect yet.

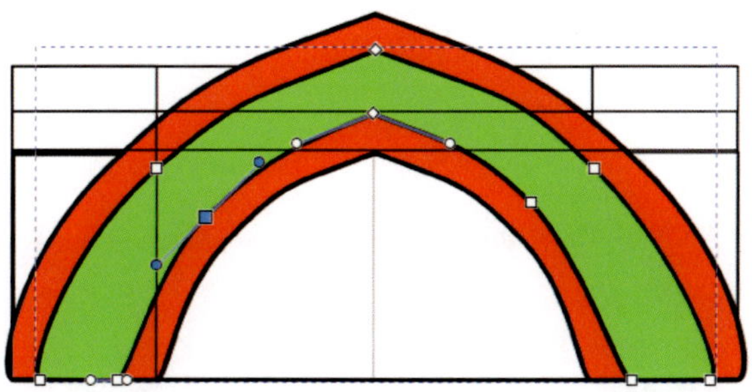

Make a copy at this stage in case it goes wrong in the next step, it's the easiest way to go back to the start.

It is all a little unevenly curved, so using the node tool, the arrow keys and the "make selected node smooth" button fiddle around, re-align nodes, fine tune the curves with their handles, press undo a lot to reverse bad moves … until you are happy with the results. When you are ready export the large curve and small curve separately.

Go into Tinkercad and import them. Align them centre and bottom.

Make a copy of the green arch and move it away. Turn the first green arch into a hole and union with the red arch.

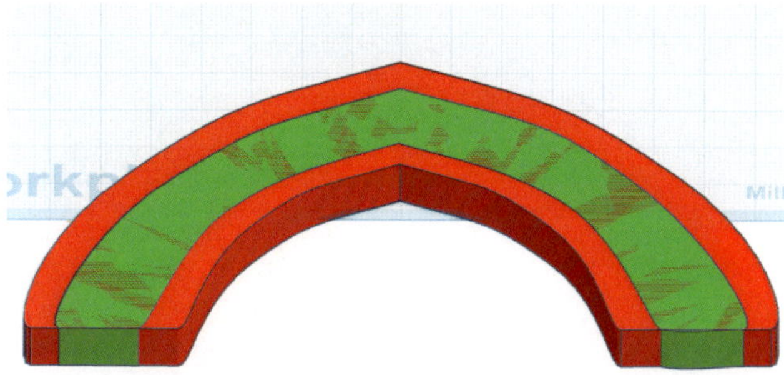

Make the red arch 23 mm high.

Bring the copy of the green arch and make it 20 mm high and align it centre, bottom with the red. Union them.

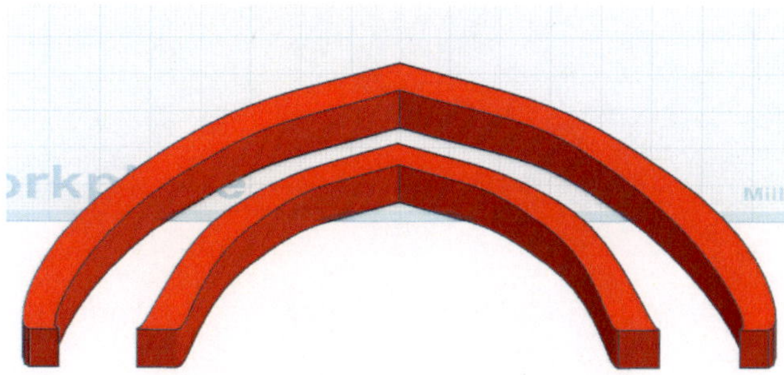

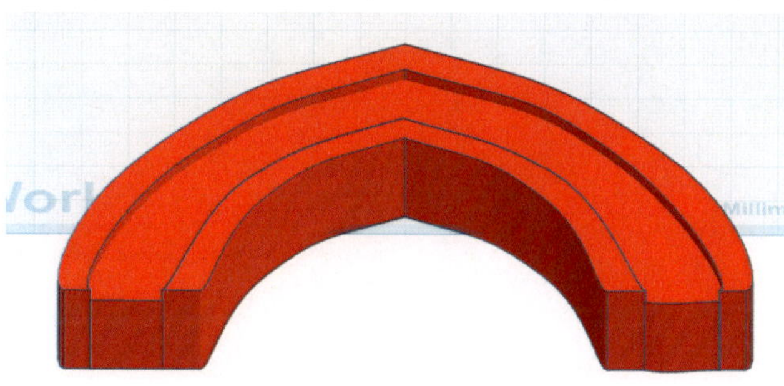

---

## Assembly

*After fitting the door and tiles (see next section) I made packing strips in black at the sides of the checkerboard - measured and printed to size.*

*I also made white edge strips, curved at the end, to meet the brickwork from the doorway - measured to the size and width of the gap.*

*The doorknob was a made from a cylinder, cone and sphere.*

# Doorstep

Normally a door will have a doorstep for the simple reason of keeping rain out of the house. If a house has a cellar it may have more than one step and the bay window may have a deeper bottom panel too. But to start simple I will do just the one.

It is also normal to have a curved piece at the front of the step. Because I am doing a mini Grand Design the step will be deep into a tiled porch. How detailed you do this, if at all, is up to you but learning the techniques is useful even if you decide against.

## Filament saving

Because the step will be deep I didn't want to print a solid lump of filament or rely on a low infill so I have hollowed the design myself with a series of boxes and a couple of cross braces. You could maybe ignore this and infill at 5% if the step isn't deep.

You can make these directly in Tinkercad or in Inkscape.

The first box is 100 mm wide by 110 mm with 96 x 106 cut out.

The next is 70 x 80 with 66 x 76 cut out.

Then 40 x 50 with 36 x 46 cut out.

The last box is 10 x 20 with 6 x 16 cut out.

Now one box 100 mm wide by 2mm at the centre horizontally and one 2 mm wide by 110 mm at the vertical centre.

Group them all and make them 14 mm high. Export to Tinkercad if you were in Inkscape.

Now to make a top for the step with a rounded front. Make a cube 100 mm wide by 110 mm and 4.5 mm high.

At the front we will add a semi cylinder of 4.5 mm wide by 100 mm and 2.5 mm high. Rotate it 90 degrees up to the face of the step.

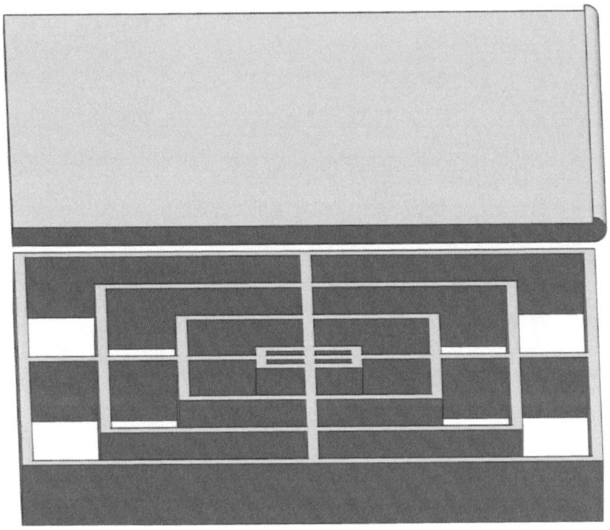

I am going to add a checkerboard tile to the top later in the tiles section, you don't have to. If you choose to make a tiled inlay we need to cut out a section to lay it in. Even on a multicolour printer it seems best to me to do this separately to reduce filament waste.

The inlay will be a hole cube 100 mm wide by 110 mm and 1.1mm high. Align it to the vertical top, centre, back of the step and union.

Add a section to go under the door which is 100 mm wide by 18 mm and 14 mm high. Box it out similar to the main step.

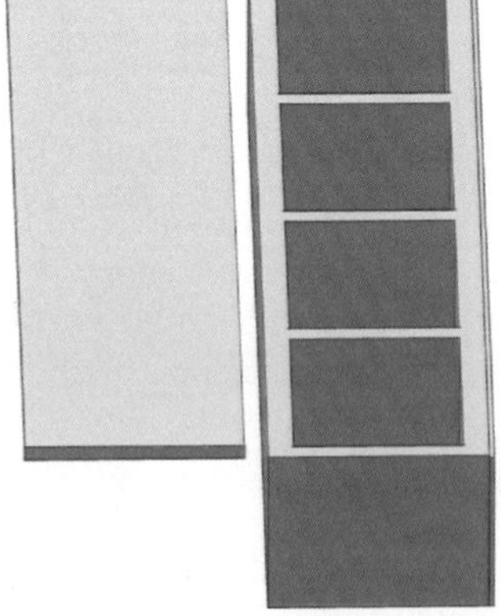

This diagram from the underneath shows the layout and also two hole cubes I am using to cut into the step slightly to put the column plinths in.

For the checkerboard inlay see the tiles section later, even with a single colour printer you can print this in black leaving holes to put white filler into.

## Leaded light

We can achieve this coloured glass effect in a number of ways. The easiest is to print or photocopy onto acetate and not use the 3D printer at all.

If you want to do it totally by hand you can use permanent markers on acetate.

If you have a four colour printer, use 3 transparent filaments and a solid black border.

Now the easy part - take the three boxes you want to fill into a design program and design.

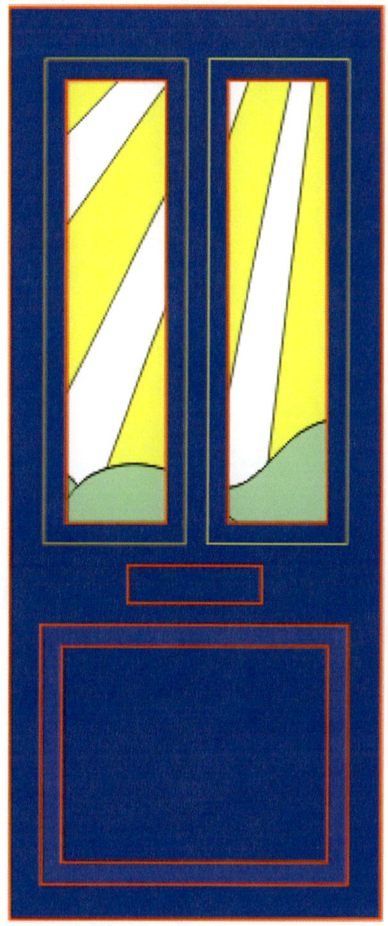

I chose the less easy route, I want a design that goes across all three window strips, so I took the boxes positioned as they will appear.

When the design is done either export it for printing in a PDF or take it out as 4 pieces in SVG - you might want to add margins around them in black too for glueing at the assembly stage (I have shown these in grey on the diagram that follows for better visibility). Also if you want spaces remember that will count as a clear filament and take your colours down to only two. My design is only yellow, green, clear and black.

## Acetate

We are going to start the design with the margins - always leave some extra to glue the acetate, and also do a fine black border where you want to align it with the window. If you make this thicker than you want visible it allows leeway for tiny misalignments. So make a 1mm black outline into 2 mm with the extra 1 mm outside the proposed visible area.

If you keep the design relatively simple it is easy to make the black joining pieces from the sections of colour. If you want something complex you will need to work out from that design in black and cut it into the other colours.

## Multicolour print

Loading up to 3 transparents and one solid black into your printer, go to the slicer and bring in the 4 SVG pieces. Choose a height of 1 mm or so - just enough to hold together as you take it off the bed.

Assign each piece to the correct colour and print.

## Assembly

If you want to enclose the acetate you can print the door in two half thickness sections, likewise print the window frame twice and sandwich the design inside them. If you are likely to see both sides of the door this is worth doing.

## Note on hinges for the door

You can make built in hinges (as in the first book Making Miniatures With 3D Printers #1) and also in the stove project later, or use metal ones screwed in from online stores that are available very cheaply. The more expensive ones appear to be just the large multipacks split down into smaller packs of hinges.

# Brickwork

There are 3 ways to print something two colour like brickwork. The simplest is to use a multicolour printer (like my Bambu A1 Combo), fill the print bed with pieces to join later and walk away. I have done this with grey and terracotta filaments.

Next if the printer isn't too old (My Sidewinder X4) it will support a change filament command in g-code and pause the print to let you switch over at the right level. E.g. you can make the mortar layer 1 mm high and tell it to switch to terracotta at 1.1 mm.

Finally for the very old (my Geeetech) I can simply make two prints - the mortar layer and a second print at Z offset 1 mm for the bricks. This is also a good way to rescue a print that failed midway - cut the object up to the level you want to resume.

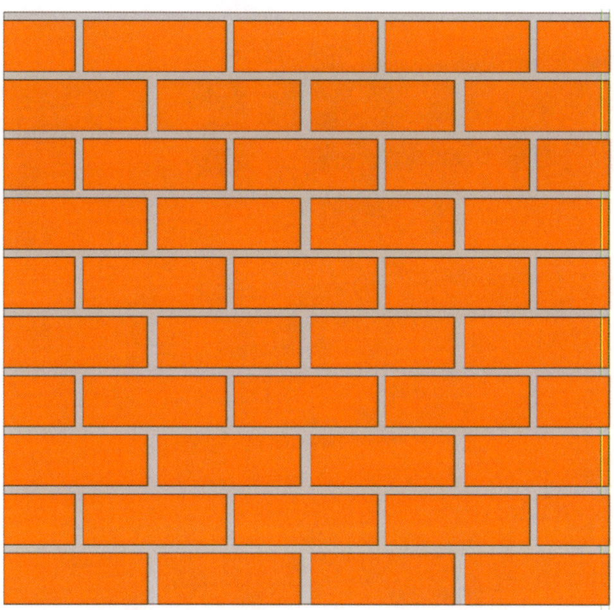

In an ideal world printers would have a 500 mm x 500 mm bed and we could print a front wall of brick with holes cut for the window and door. Mostly we don't have that and stop just short of 250 mm / 10″. So we have to work in pieces, and with brickwork interlocking that requires a quick bit of thought about what size pieces to use and where to put the joins.

Also if you want to go around the corner you will have to print L shaped bricks. The corner pieces should be as thick as mortar plus brick combined, in my case 2 mm. I have chosen an interlocking scheme so I can add around a corner or on the straight. This way I can place pieces together in Tinkercad and cut out holes, exporting them to print as many as I can fit on a bed. It is likely to be two or more prints to print a simple one storey frontage.

I have chosen to make pieces the scale equivalent of 3ft wide with interlocks and corner option and 3ft high, aiming for a 15 foot by 9 foot front giving 5x3 pieces. The window and door will carve large chunks out of this so reducing the amount of print to about 2/3rds of that.

Standard brick size for the Victorian era is 228 mm / 9″ by 114 mm / 4.5″ and 75 mm / 3″ high. Each era tends to have a slightly different brick size, though will normally be twice as long as deep. I have gone with a Victorian size as a likely option for the style of frontage I am building. This makes a brick about 18 mm by 9 mm and 6.25 mm high.

The style for the flat interlock is simple enough, at once size cut off half a brick and at the other let the brick at the other side run half over the join. The mortar needs to be designed not to get in the way.

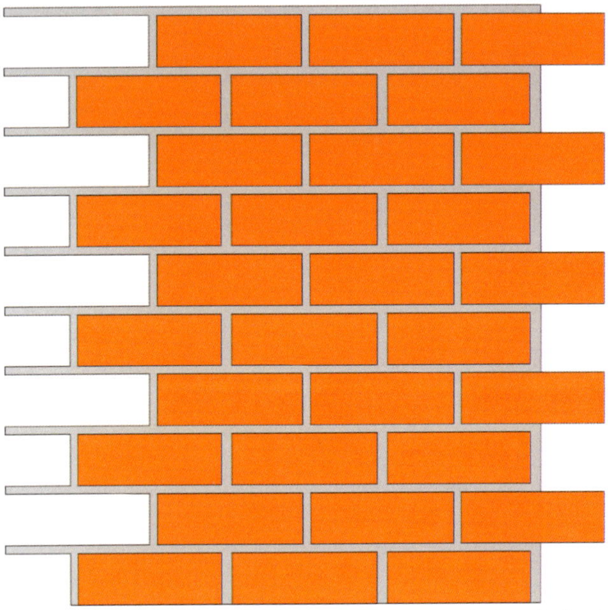

The style for a corner interlock needs both half and full brick holes. The corner brick will fit one way or the other at the other side of the wall. If you intend to carry the wall on around there we also need a piece to run back into the flat joining pieces.

Once we have arranged bricks spaced at 1 mm apart top and side, the grey (mortar) pieces can be made by cutting the bricks out out of them.

Export the mortar and brick pieces separately.

For two colour printing we need to add a Z marker to allow us to raise the brick piece off

the mortar as by default most slicers place objects down to the bed otherwise. So I make a 3 mm cylinder of 0.1mm high and group it to the brickwork. You only need one of these per print bed if you export all the pieces together but I find it is easier to have the flexibility of adding it to every type of piece.

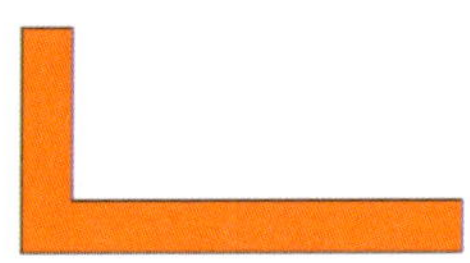

*corner brick*

*The brickwork was cut to size and assembled in Tinkercad*

# Stove / Range

## These projects teaches you how to:

- *Make complex shapes from simple pieces*
- *Add hinges and latches*

# Stove / Range

This project is an old fashioned stove or range with wood surround, hinged doors and back tiles.

It's a box with a top full of varied sized holes with covers to allow a pan to sit inside or a kettle on top. There are compartments inside the stove and a firebox near the top centre.

One large compartment at the left is split into two levels with a grill and a large opening at the top.

The centre is two level with another large opening at the top, and the right is one large compartment with two small openings at the top.

All these openings have matching hot plates with a small recess to fit in the holes.

There are doors and small panels at the front. The upper centre is an open fire with grilles across.

Around the stove is a wooden fireplace.

Behind the stove inside the chimney are tiled panels.

At the bottom front is a small, angled, grille piece.

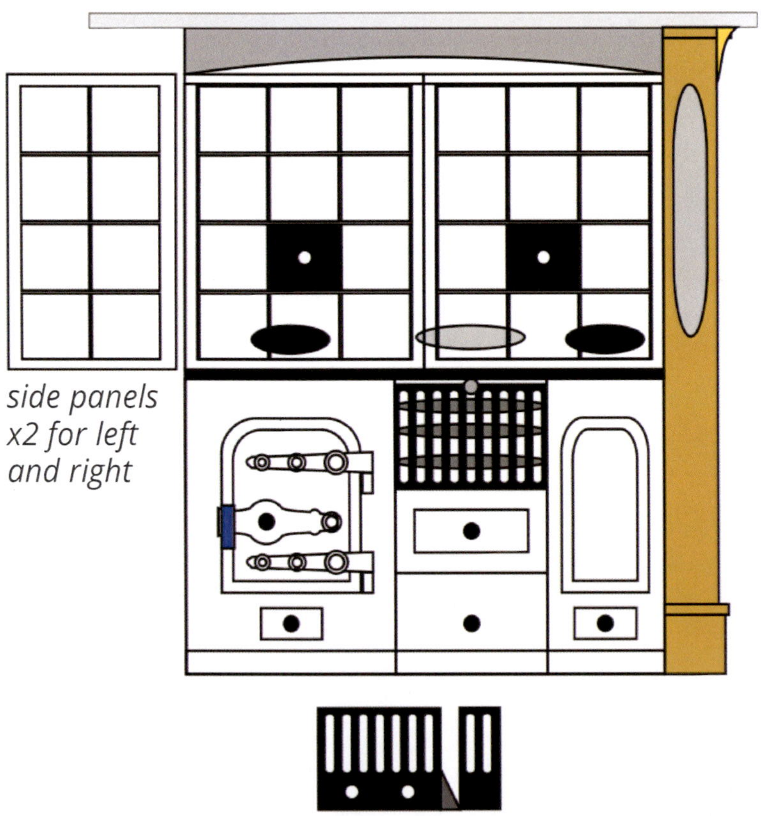

## Planning

As always I started with a diagram of what I was going to make. This can be a sketch by hand, but it is a lot easier to draw it up in Inkscape and transfer pieces directly out from there.

For the stove surround I elongated the legs from the one we made in the first fire surround project, to leave an opening 127 mm wide by 167.5 mm. I also added a curved piece of wood under the mantlepiece.

The stove will be 127 mm wide by 79 mm and a tiled back above it of 127 mm by 76.5 mm.

The tiling will be in four panels - two of 63.5 mm wide at the back with 3x4 tiles in each and two of 45 mm wide at the sides with 2x4 tiles each.

For the metalwork of the door I found photos on the internet and made approximations that I thought would print and work well. The hinges are quite clunky on most of them which is good as at 12th scale and 3D printed they would be very fragile otherwise. I checked around Angie's craft room and we found some very stiff wire of about 1 mm that I could cut and insert into the hinges after printing.

A lot of the door furniture I decided would be just for show, with only one opening main door at the left and opening hot plates on top.

I also decided I wanted parts of the furniture to be in a metallic filament to highlight against the black of the main body.

My last bit of planning was the tiles at the back, I used plain white which was unusual in Victorian times. Later in this section we will do encaustic and "painted" tiles and I think a simple painted-effect white tile with a colour design will be the best representation of what an old stove would have, and is quite simple on a multicolour printer. With a single colour machine you might need to make a stamp for them and use actual paint.

## The shell

Let's start with the main body of the stove.

First of all a backplate made from a cube of 127 mm wide by 76.5 mm and 3 mm high.

Make 4 cubes of 3 mm wide by 76.5 mm and 45 mm high. These will be the sides and dividers. Align one to each side of the backplate.

Make a cube of 127 mm wide by 6 mm and 45 mm high for the base.

Make a cube of 55.5x3x45 and align it to the top left backplate

Make a cube of 31x3x45 and align it to the top right backplate

We can now position the dividers. Align one divider to the right of the top left piece, and one to the left of the top right piece.

Make a cube of 40.5x3x45 and align it centre, centre to the backplate. Now align it to one of the centre uprights.

You can bundle or union at this point to stop things getting moved around.

Now for a shelf in the left compartment. The doors and shelves are easiest in Inkscape, though still possible with patience in Tinkercad it will take more effort.

The shelf is 49.5 mm wide by 45 mm. It has 9 grille holes evenly spaced inside it.

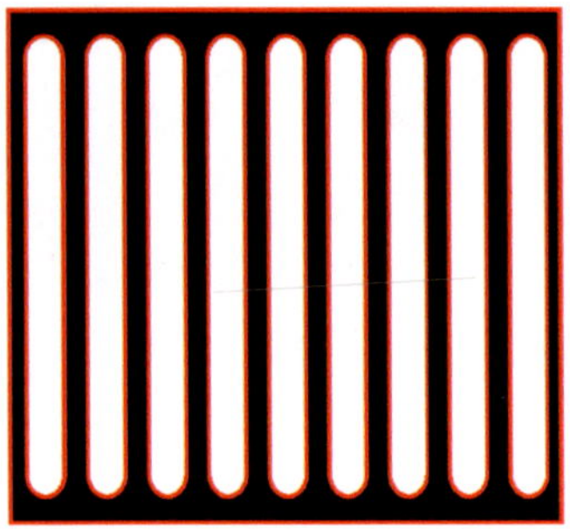

You can make these square or can curve off the ends as I have done (a quick way is to add circles to the end of each and union them). They are 3.5 mm wide by 40.5 mm. Make 8 copies. Move them to vertical centre, move one near the left edge and one near the right - don't worry about exactness yet.

Distribute the copies from the Align and Distribute tab and group them all, You can now align the group to horizontal centre of the box, and Path -> Difference to make the grill.

Export and take into Tinkercad, make it 3 mm high. Rotate it on end and align it into the shell, level with the centre shelf but between the two divider walls.

## Doors, hinges etc.

To make the rounded topped doors in Inkscape, make a box of 37 mm wide by 35 mm. On top, centre add another box of 17.5 mm by 10 mm.

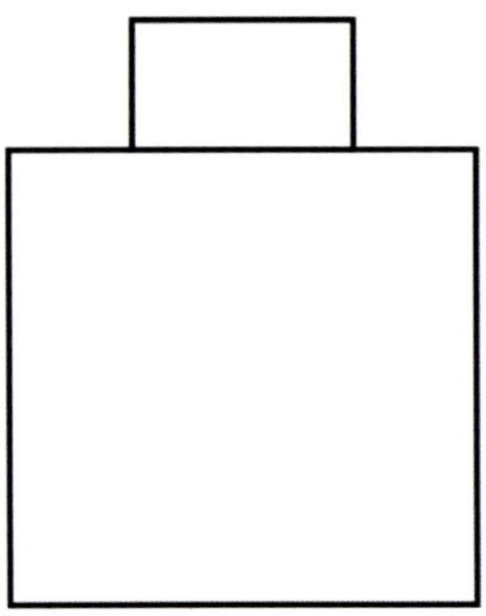

Now draw a circle of 20 mm and align it top right to the two boxes.

Copy it to top left, and union them all.

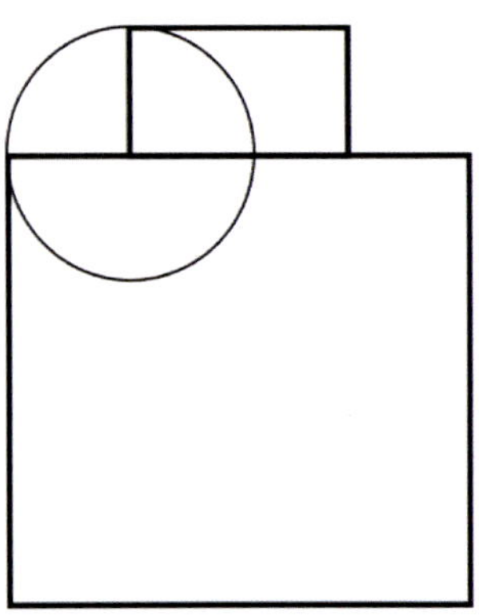

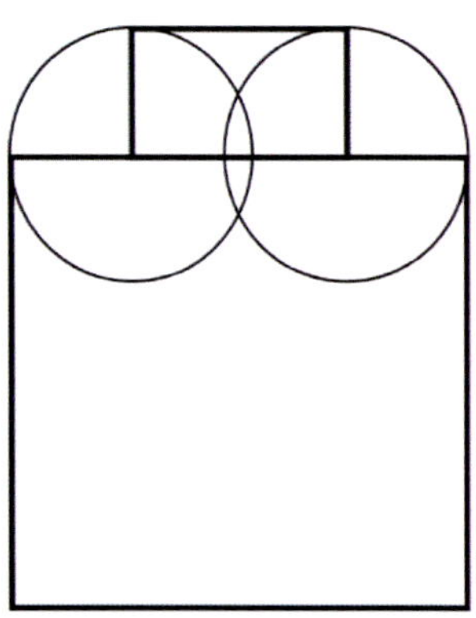

Make a copy of the union and scale it to 85%.

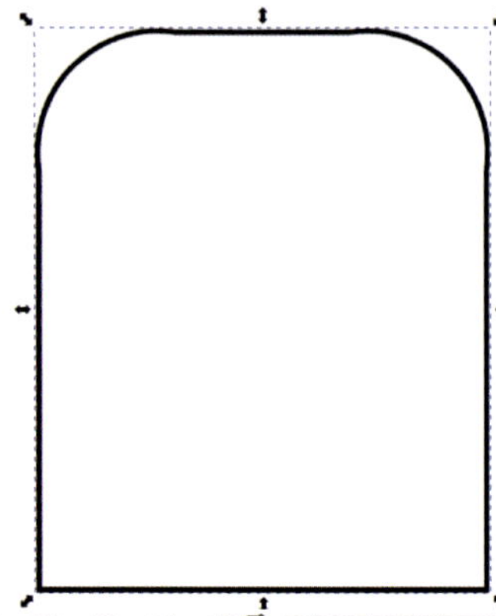

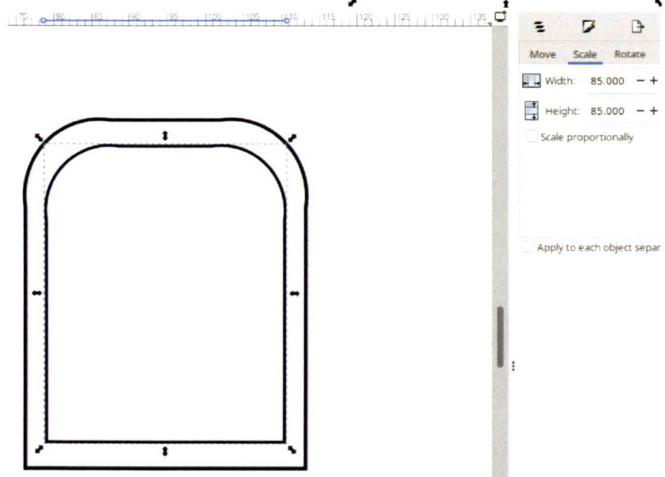


Next we'll make a hinge. Start with a box of 31 mm wide by 5.5 mm. In Preferences _. Behaviour -> Steps make arrow keys move by 0.5 mm.

Click on the box and Path -> Object To Path. Now with the node tool click on top left and move it down 4 arrow keys (2 mm), then on bottom left and move it up the same amount.

Make circles of 6 mm, 4.5 mm and 4 mm and spread them evenly across the box.

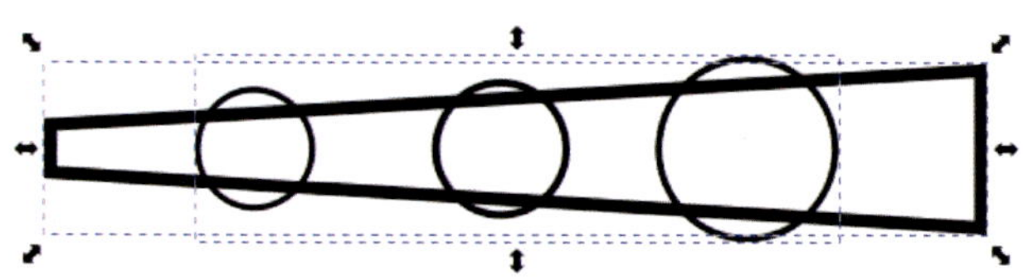

Draw circles of 4 mm, 2.5 mm and 2 mm inside these circles (these are markers for studs).

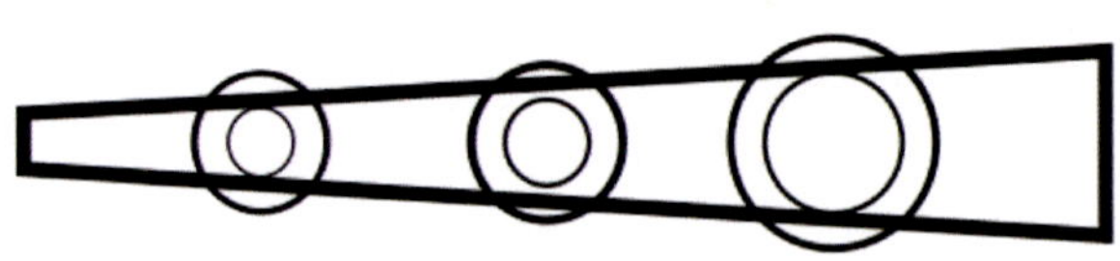

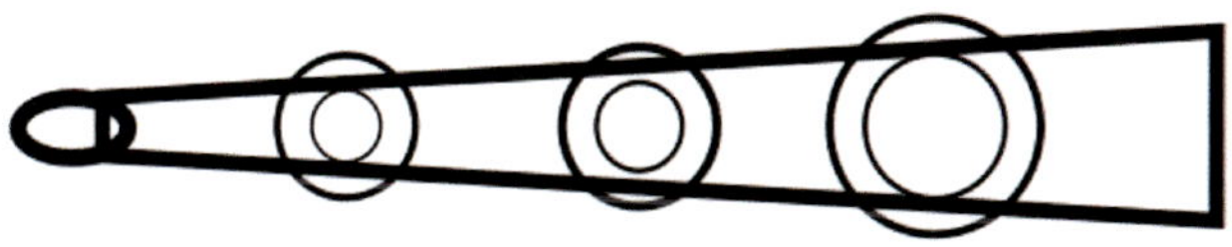

At the small end draw an oval of 3.5 mm wide by 2 mm and move it in a little so the curve meets the square end.

Union everything except the 3 inner circles.

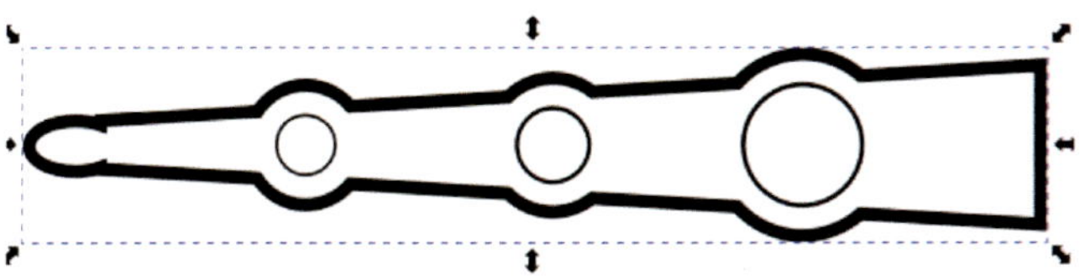

For the latch draw a box of 33 mm by 7 mm. Draw a circle of 11 mm and put it roughly centre.

Draw a circle of 5 mm and put it halfway across the right end.

Draw an oval of 4.5 mm wide by 6 mm and slightly overlap the 5 mm circle inside the box.

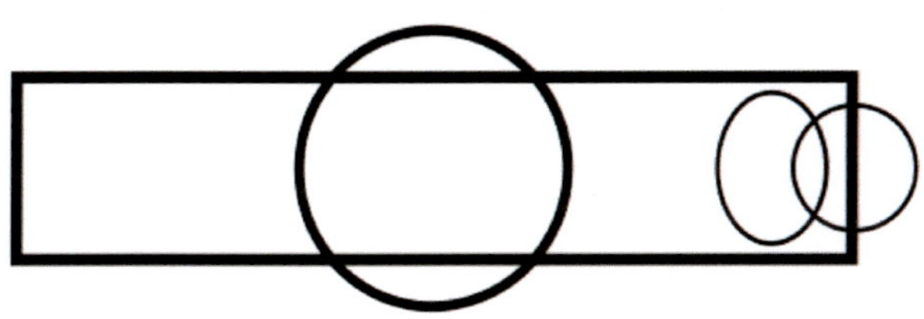

Now turn the box to a path Path -> Object to Path.

Click on the top right node and move it down 2.5 mm, and move the bottom right up 2.5 mm.

Centre a 3 mm circle on the large circle (this is a marker for the handle), and 2.5 mm one on the small circle (this is a marker for the hinge).

Union everything except the two small circles.

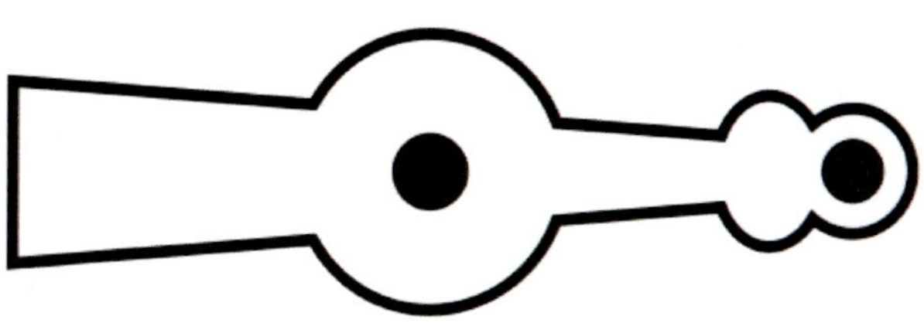

This is a good-enough part as it is, but why not play with the node handles and make it even better?

Select all the nodes and click "Add curve handles". Now manipulate the straight sections to curve gracefully into the circle sections. Undo is your friend, but it's best to make a copy of it all first.

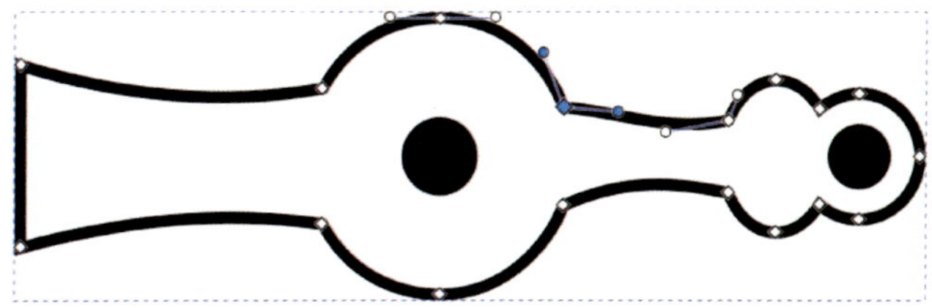

Finally draw a box 3 mm wide by 11 mm near the top left of the latch. This is a reminder to make a hook for it to sit in.

For the front panel draw a box 127 mm wide by 76.5 mm and align it to the shell.

Cut a square out of it at the centre of 40 mm by 28.5 mm, aligned inside the inner supports.

Arrange the two door parts to the centre of the left compartment and 21 mm up from the floor.

Copy the inner door and cut it from the front panel.

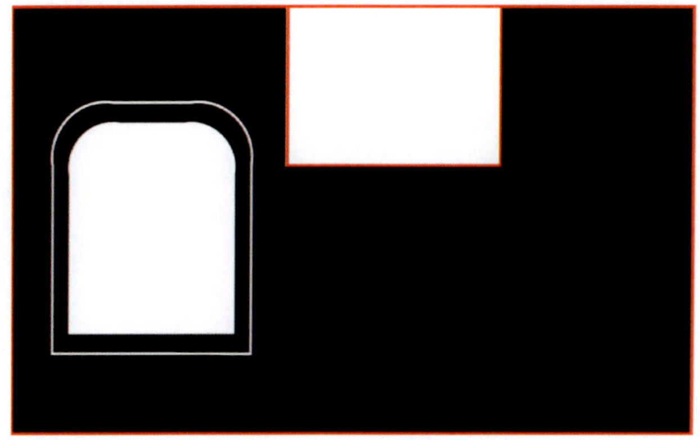

I have added four boxes for front panels wherever I feel they look good, with markers at their centres for handles (see the diagrams at the start of the project). These will be printed on the front and not open but it looks better with them there. Feel free to make holes and put removable boxes or doors there.

Above the front draw a box of 127 mm wide by 46.5 mm. This will be the worktop.

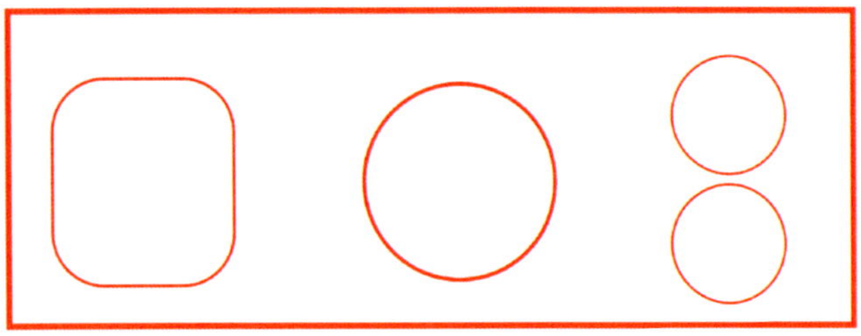

Draw a square 27 mm wide by 30 mm. In Inkscape, if you don't convert to a path you can use the node handles to round off the corners. So pull one of them in about a quarter of the width to round them.

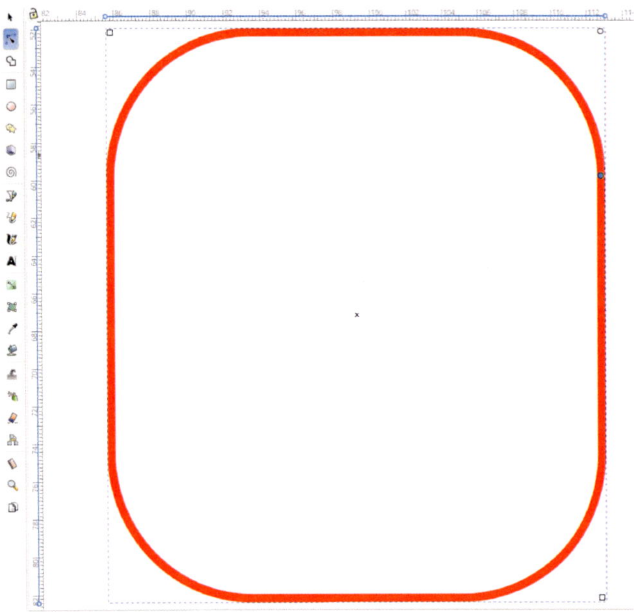

In between the centre supports draw a circle of 29 mm.

At the right hand draw 2 circles of 17.5 mm.

Align them all to look good but also to be at least 3 mm from the back and sides as we will be building up there. Select the top and each hole one by one and click Path -< Difference.

We now have the hot plate holes in the counter top.

We can now ungroup the shell and take away the top right and top left pieces. We will use the top we created twice in Tinkercad to make alignment easier. First we will replace the top pieces we just removed, then we will cut a hole in it for the centre and finally put the complete top across the top of the shell. This will leave holes reaching inside the shell.

## The surround

A quick repurposing of the fire surround and mantle we made in the first project will do for this. Take a copy and ungroup it.

Increase the leg to 167.5 mm from 99 mm. Stretch the oval indent with and place it a little lower - 15 mm from the top to be roughly centred on the tiled pieces we make later.

Make the mantle 177 mm by wide 4 mm deep.

Make a cube of 127 mm by 12 mm and 10 mm high for the arch.

Make a hole cylinder of 127 mm by 12 mm and 10 mm high.

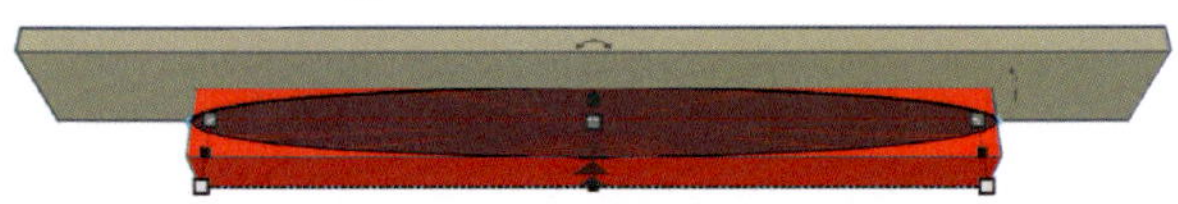

Align the mantle, arch and hole to centre, align the hole and support vertical centre.

Move the hole 6 mm down and union it to the support.

Move the legs to either side of the support. This should leave a gap of 127 mm wide for the stove.

## The tiled back and side walls

For the tiled side of the recess make a cube of 42.5 mm wide by 76.5 mm and 2.25 high for the backing.

Make a hole cube of 38.5 mm by 72 mm.

Align it vertical centre and to the right of the backing, them move it left by 1 mm and union. I print two of this piece, one mirrored L-R, in black and slot in the tiled pieces.

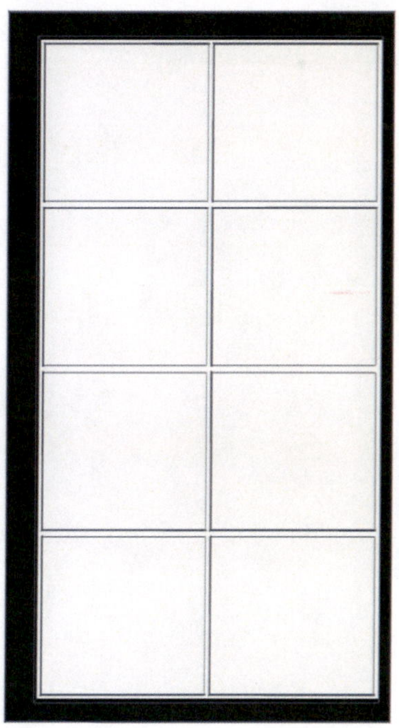

For the tile backing make a cube of 38.2 mm by 71.7 mm and 1 mm high.

For the tile make a cube of 18.25 by 17 mm and 2 mm high.

Make 7 copies and arrange them all spaced 0.5 mm from each other and the edge of the backing.

Union them all. We now have a tiled piece that I printed in white.

We will make two identical panels for the back of the stove. Optionally we can replace one of the tiles near the lower centre with a fake door for access to the chimney.

Make a cube of 63.5 x 76 x 2.25 for the backing

Make a hole cube of 57 x 72, align centre, centre and union.

Make a cube of 56.7 x 71.7 for the tile backing and copy 4 rows of 3 tiles into it.

If you are making a door to the chimney take a tile out to leave space.

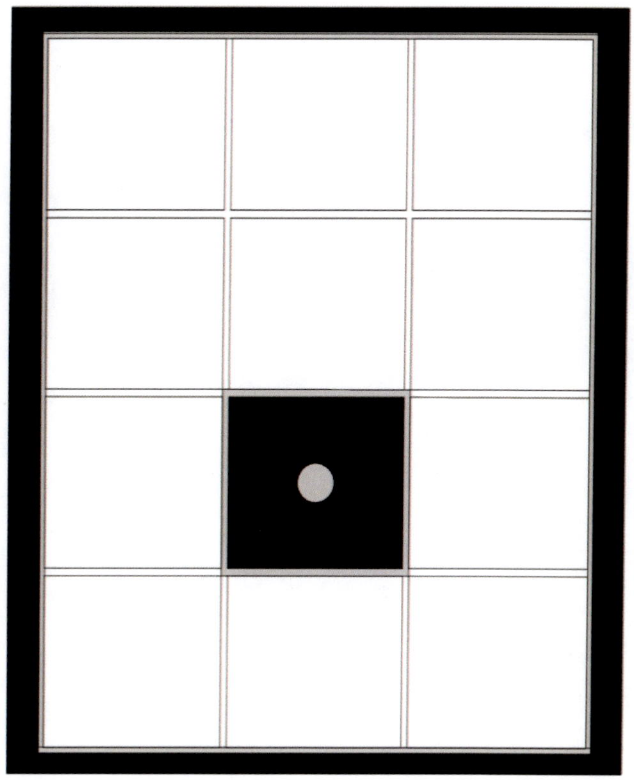

Print two identical pieces and tiles.

# Assembly

We've made a lot of pieces and we still have a handful more to go. We need to make hot plates for the holes in the stove top, we need to make hinge posts for the door and we need to turn the grill piece into a free standing item at the floor of the stove.

## Hot plates

The lids are a composite of a slightly smaller than the hole piece (to rest in the hole) and a slightly larger then the hole piece (to cover it). Export the three shapes and sizes of hole individually from Inkscape. Simply copy each shape and make it 1 mm smaller all around then make the copy 2 mm bigger all around. Make the large copy 1.25 mm high and the small copy 2.5 mm high and centre them.

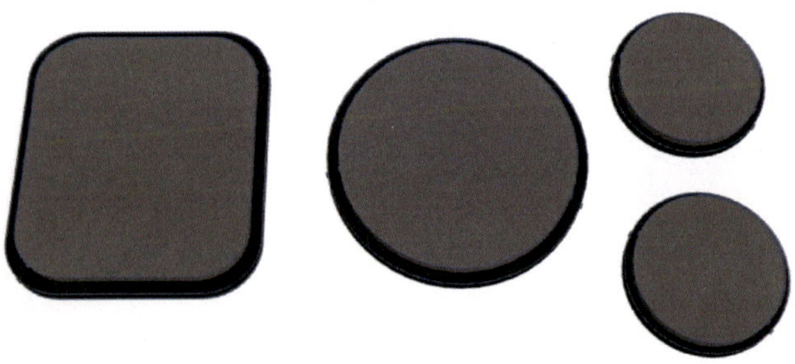

## Hinges and latch

We have design choices to make here. If you have a metallic filament you can make the hinges and latch out of them. If you have a multicolour printer you can build the metallic pieces into the black door in one. You can build the hinges into the door all in black and just use the metallic for the door latch.

The circles in the hinges should be raised slightly above the hinge to look like large rivets.

At the side of the hinges we need to put a cylinder with a hole to allow it to swivel on a wire, and beneath that a cylinder with a flat base to fix to the front of the stove.

In the same way we can choose whether the fake doors are metallic or built in to the black

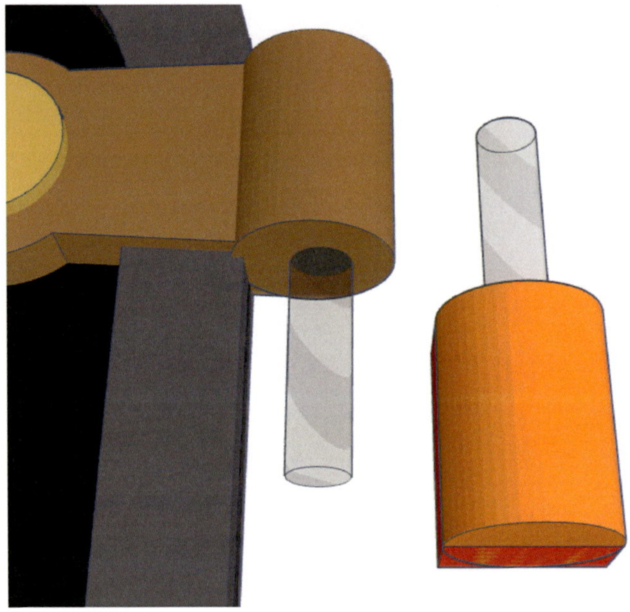

front. As long as they are raised they will look good.

The hinge is cylinder of 4 mm and 5 mm high rotated 90 degrees and moved to touch the edge of hinge, then centre it.

Make a hinge hole with a cylinder of 1.5 mm, centred in all directions and then moved 1.5 mm from the top of the cylinder.

The hinge body is a copy plus a cube of 4 mm wide by 5 mm and 2 mm high.

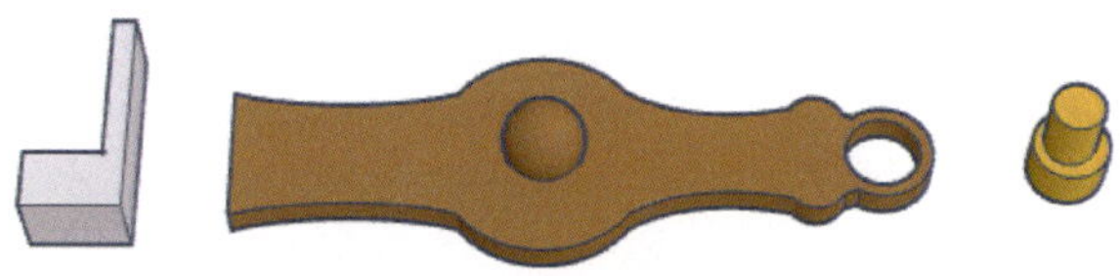

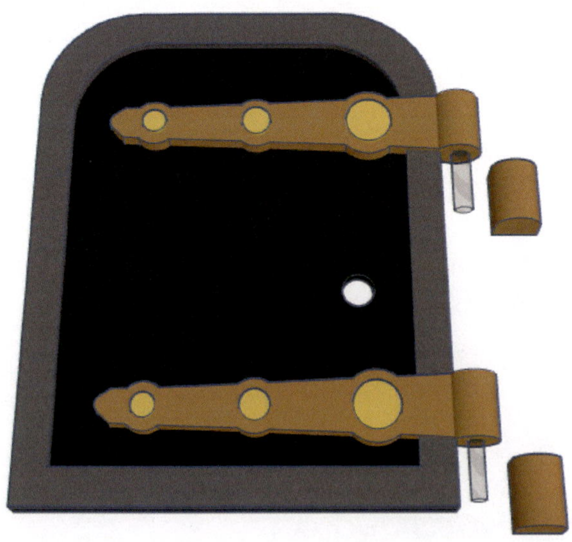

Bundle and arrange two hinges by eye (vertically and horizontally) on the door, deciding if you are bedding it into the black door in one print or cutting it out of the black door (as a hole) to print in metal.

Position the hinge body on the black front where it will perfectly receive the hinge pin (1 mm wire added later).

Position the door latch on the door between the hinges and use the leftmost circle to cut a hole for the pin allowing it to swivel.

The pin is a cylinder of 4 mm and 2 mm high with a cylinder of 2.5 mm and 5 mm high.

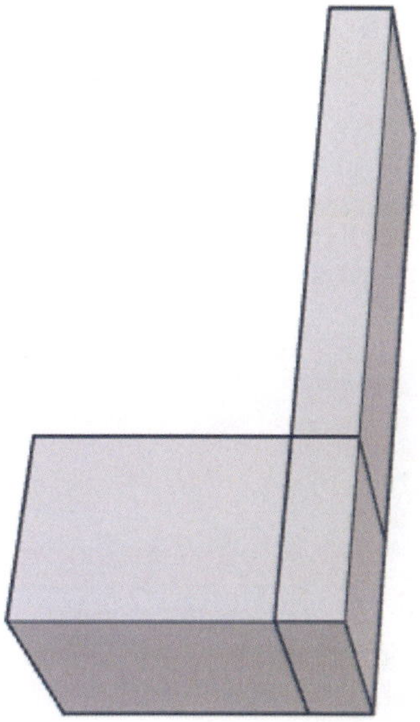

The catch for the latch is a cube of 5 mm by 3 mm and 3.5 mm high with a cube of 11 mm by 1 mm and 5 mm high. The catch glues onto the front body and is best positioned after mounting the latch.

The second door at the right is false and just a copy of the first door "squashed" to size.

To each of the false rectangular door panels add a knob, e.g. a sphere aligned to centre.

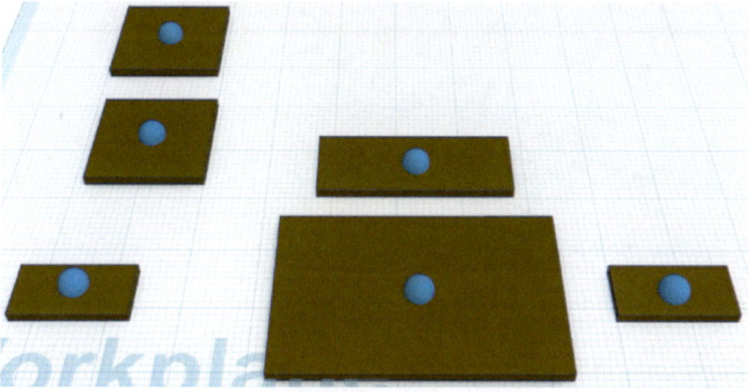

There's a little grille at the bottom of the stove which is just the grille we used as the shelf re-adapted.

Add a wedge of 4 mm deep and 10 mm high the full length along the front. Take a copy of the grille and wedge and use a hole cube to cut it down to just two holes. Angle them 45 degrees and move to touch the front grille. If you need to cut off any excess, angle a hole cube and union with the wedge. Now copy and mirror the side piece to the other side.

Carefully choosing which piece will be which colour, export the pieces and print. They should easily glue into place though I often use a little jig I made in the first book which is two thick pieces of wood at a right angle on a board. This allows me to screw scrap wood into place to hold joints until they dry.

*I made a padding strip the same width as the board I attached the stove to, so it fitted across the top of the tiled sections at the front. I also scaled the sloped chimney piece from the fireplace to fit behind the board at the back.*

# Kitchen unit

## This project teaches you how to:

- *Make pieces with multiple angles*
- *Add texture and grooves*
- *Split pieces that are too big to print*

# Kitchen unit

Here we make a kitchen unit with a built-in tap and sink, draining board and glass cupboards above. Of course there are diagrams, it's so much easier for me that way.

I haven't made the bottom cupboards to open - they are just fake panels and I only intend to lightly stick the upper glass doors on so I can go in and re-arrange crockery. If you want to, you can separate them and put in some hinges. I will be putting in side and centre posts for this anyway as it wouldn't look right without.

## The shell

the main body is 144.5 mm wide by 203 mm. This would be near to the ceiling in most houses. I left a little for decorative coving at the corner of the ceiling.

I did this in Inkscape but you can work directly in Tinkercad. Make an inner box 138.5 mm by 188 mm, centre it and make it 4.5 mm from the top and 10.5 mm from the bottom.

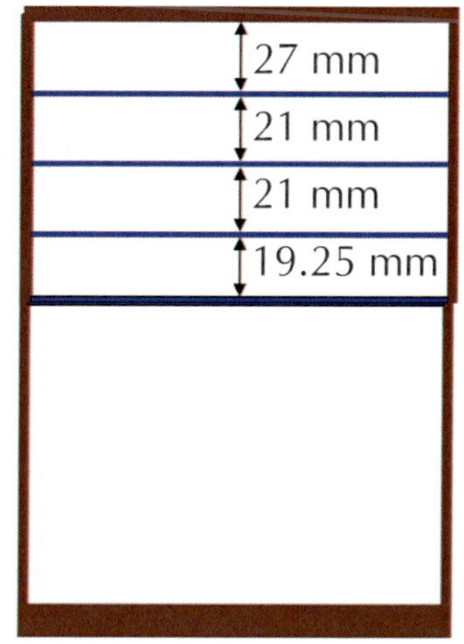


I didn't space the shelves evenly, you can, I will just say what I made here.

Draw a marker 27 mm aligned down from the outer top of the cabinet. Make a shelf 139 mm wide by 1.7 mm and place it below.

Draw another marker of 21 mm below and copy a shelf below that.

Repeat, then do a final marker of 19.25 mm and copy a shelf below that. Delete the markers and centre all the shelves inside the shell.

Now for the lower shell (note here that if you want the doors to open you will need to add two uprights for them).

Make a box of 139 mm by 56.75 mm for the lower shell.

At the top centre another box of 58 mm by 20 mm.

Vertically centre a box of 36.25 mm by 52.75 mm, move it to 2 mm in from the left and move a copy 2 mm in from the right.

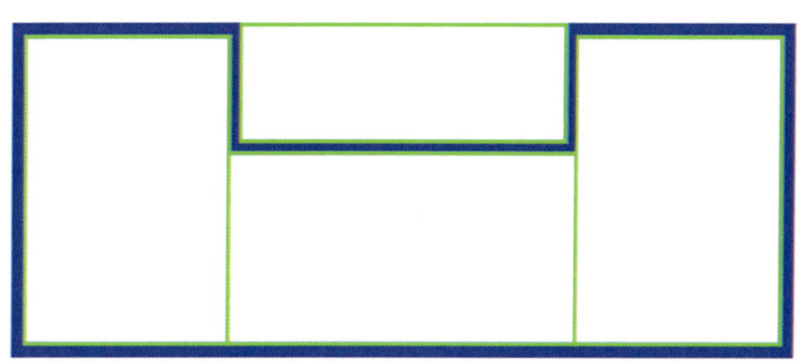

Draw a box of 62 mm by 32.75 mm and centre it, then move it to 2 mm from the bottom.

Use Path->Difference to remove each box from the lower shell.

## Planking for the back

Draw a box of 144.5 mm by 155.5 mm as a guide.

Draw a box of 5 mm by 155.5 mm. Align it left to the guide box.

Make a copy and align it right.

Now make 24 more copies and use Align and Distribute -> Distribute -> Distribute horizontally with even spacing. This will make the backdrop. You can delete the first guide box. Group the planks.

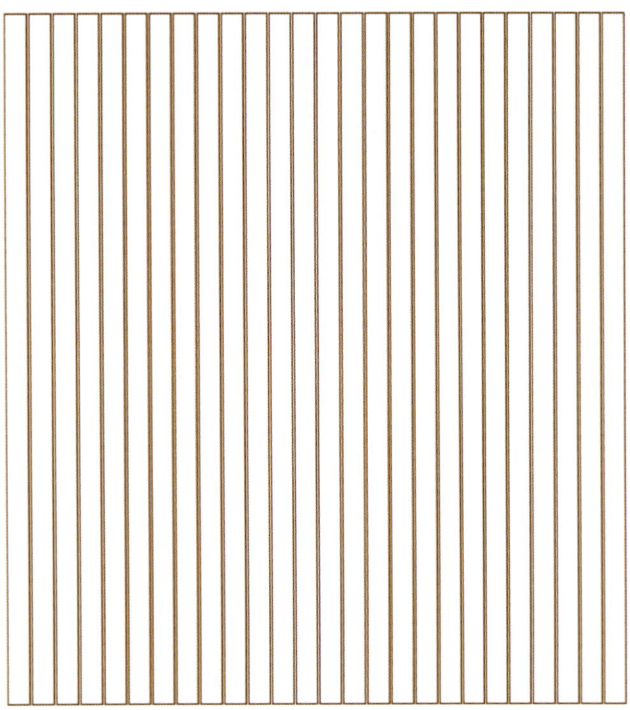

Export the outer shell, shelves, planks and lower shell to SVGs and bring in to Tinkercad.

Make the outer shell 46 mm high and the inner pieces 43 mm high. I have made the shell inners - lower part and shelves - recessed by 3 mm so that the fronts click on perfectly flush with the outer shell.

Make the planks 3 mm high. Centre them to the top of the outer shell.

Draw a cube of 144.5 mm wide by 203 mm and make it 2 mm high, centre it to the outer shell.

Centre the lower shell and move it 10.5 mm from the bottom.

In the screenshot you can see I added a centre post to the upper shelves of 2 mm wide and a ledge of 2 mm deep just under the top edge. These are to make it easier to stick the upper doors to.

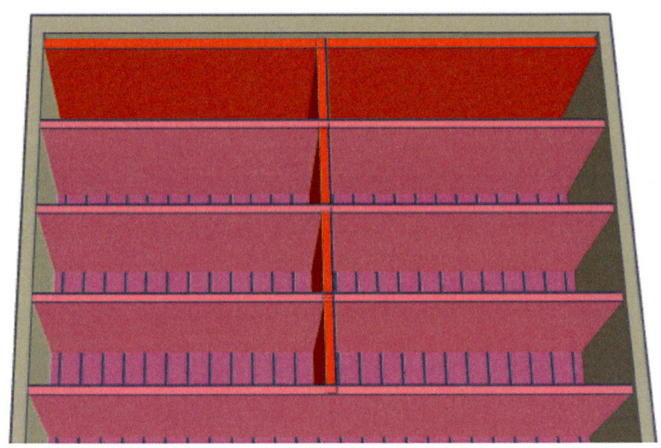

Union everything. If you have a large bed you can export this off for printing now, if not you need to split it in two. You can do this here in Tinkercad or in the slicer.

The quickest way in Tinkercad is to make a hole cube for the top half and one for the bottom. Union one then export. Ungroup then repeat. I have to as my maximum print size is just over 24 cm (under 10″).

In Prusaslicer there is a cutting tool but it is more complex than the hole cube method.

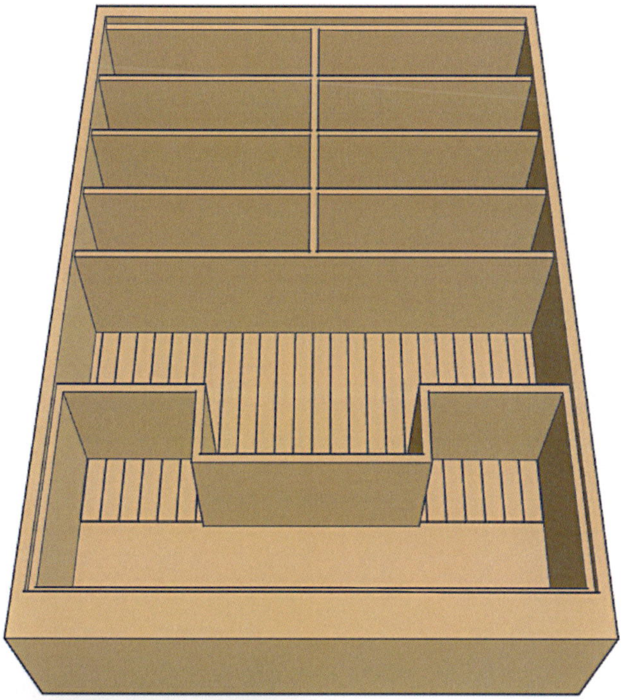

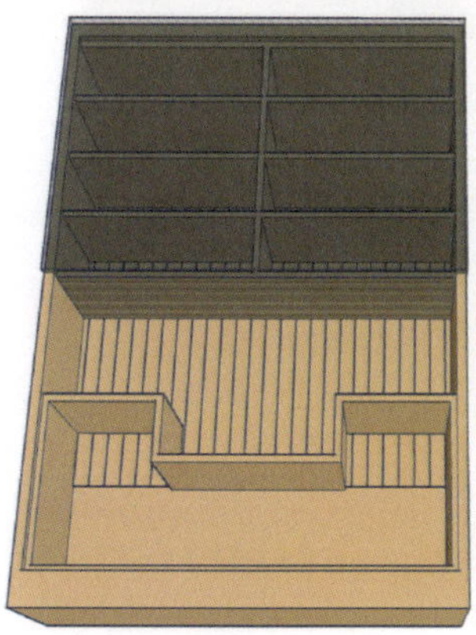

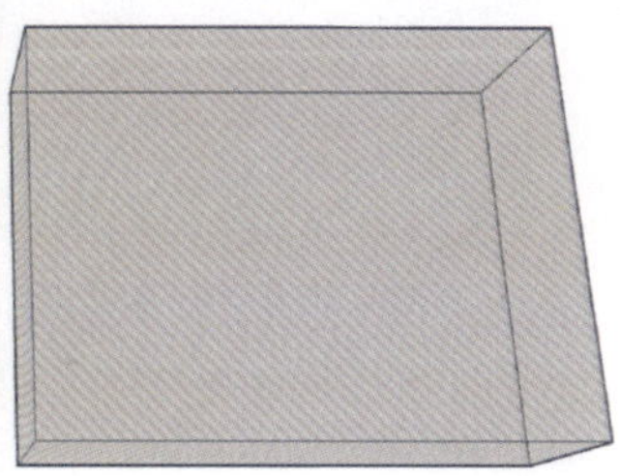

## The upper front

This is fairly simple, four doors held together by 3 uprights.

The door is a box of 32.5 mm by 90.25 mm with a hole of with a hole of 23.5 mm by 80.25 mm centred in it.

Make a box of 10.5 mm by 90.25 mm for an upright and align it so that the left half touches the right edge of the hole. Copy the door and align it so that the upright goes up to the right edge of the hole.

Make a box of 15.75 mm by 90.25 mm and align it to touch the right edge of the second door.

Copy the two doors and first upright over to the other side of the centre upright, aligning to the hole left edge again.

We now have three uprights to export and make 2 mm high in Tinkercad.

Export the doors and make them 3 mm high.

Union the doors and uprights, export and print in grey-green.

## Lower front

Make a box of 138.7 mm by 56.75 mm for the lower front. Cut a hole out of it top, centre of 38 mm by 20 mm.

We will add fake door and drawers to it (unless you want to go ahead and make them real).

The drawer fronts are a box of 39.25 mm by 10 mm.

The large fake door is a box of 38 mm by 42.5 mm with a hole cut out of 28 mm by 32.5 mm.

The small fake door is 30 mm by 34.5 mm with a hole of 19 mm by 24.5 mm.

Align them all to look good and take into account that if you want to put simulated hinges in place you need to leave places for them, e.g. the centre post for the small doors.

Export into Tinkercad and make the lower front 2 mm high and the fake pieces 3 mm high.

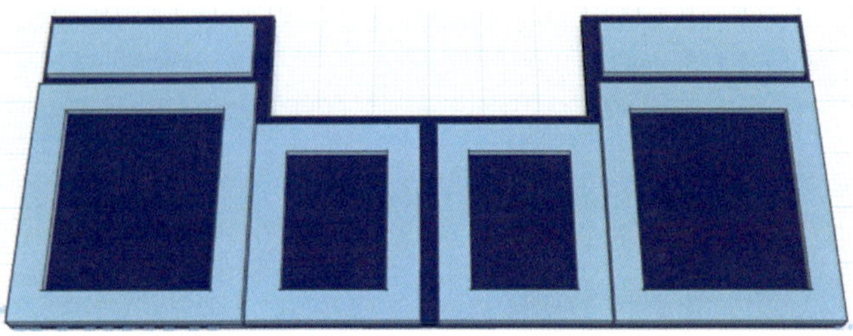

You can add hinges or door knobs to taste, small metallic ones are  available cheap online or you can print something small in metallic filament to simulate them if you don't mind them not opening.

You can now print the four (five if split) main pieces - I printed the shell in light wood, the draining board in dark and the unit fronts in grey-green.  I added acetate for the glass in the upper front.

## Draining board

The draining board is a surprisingly complex shape with run-off into the sink in all directions from slopes and cut lines.

Start with a box of 138.5 mm by 46 mm.  Into the centre, bottom cut a hole of 35 mm by 40 mm.  This is the basic board.  Copy it and paste.

Make a bigger box, e.g. 150 mm by 60 mm and Path-> Difference the basic board from it. We will need this later to tidy up in Tinkercad.

At the back we will add a plateau for the tap and the back of the draining board.

Make a box of 138.5 mm by 6 mm and align it across the top of the board.

Make another box of 40.25 mm by 6 mm and align top left.

Click Path -> Object to Path and click on the node tool.  Move the top right node in 6mm.

Copy and mirror it horizontally and align to the right.

Copy and paste both pieces. Individually Path -> Difference them to the long top bar.

We now have 3 angled pieces along the top.

Make a box of 40 mm by 46 mm and align it top, left this will be the left board.

Copy the centre angled piece and Path -> Difference it to the box.

Copy, mirror and align the new box to the right.

For draining lines make a box of 38.5 mm by 0.5 mm.  Align it to the bottom of the centre angled piece and to the right of the left board.

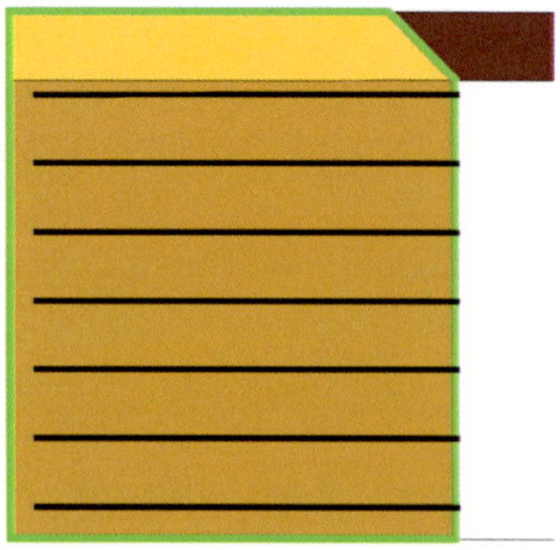

Duplicate the line down by 5.5 mm until you are near the bottom of the board.

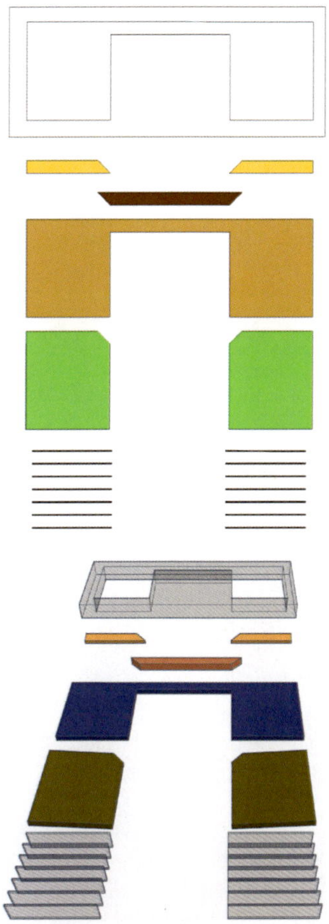

It will be 7 pieces in total.

Group the lines and copy them and align left to the other side board.

We are ready to export these pieces out to SVGs and into Tinkercad.

Make the basic board 3 mm high.

Make the left and right top angled pieces 3.5 mm high.

Make the centre angled piece (for the tap) 4.5 mm high.

Union all four pieces.

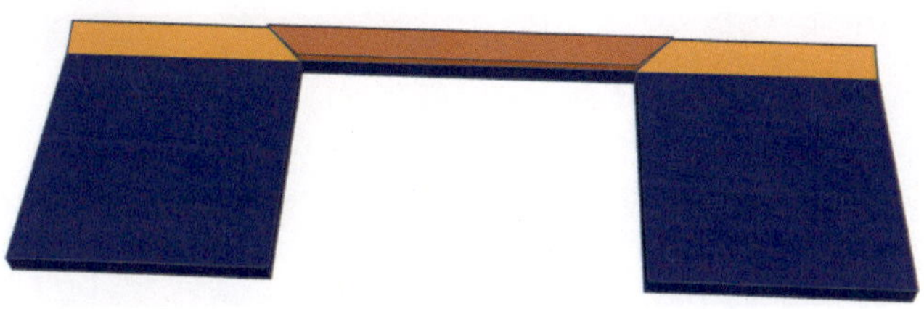

Turn the lines into holes.

Make the left and right board 4.5 mm high.

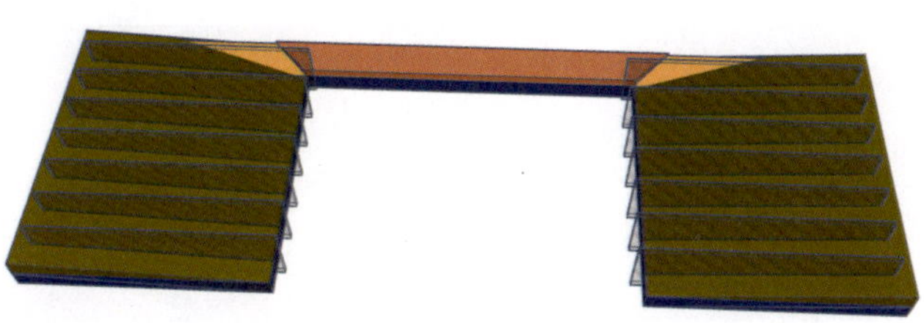

Now angle them inwards about 6 degrees or so to raise them at the back and just cover the basic board at the front.

Union the lines with them.

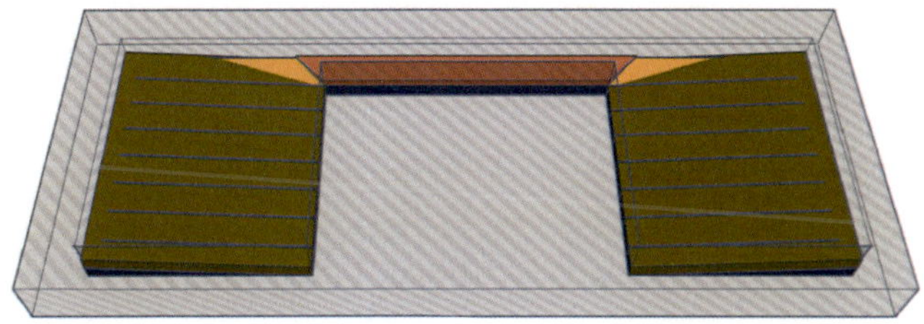

Finally make the template into a hole and union it with everything - this will preserve the exact shape of the piece even after rotations and additions.

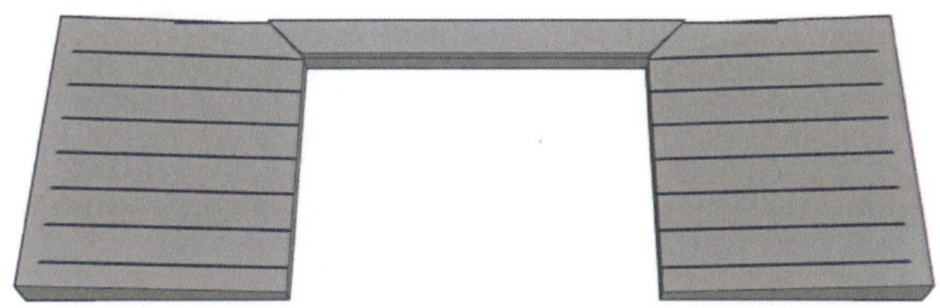

*The crockery was a free download just to fill the display.*

Plate 3CM by DiamondStarr on Thingiverse:
**www.thingiverse.com/thing:6000502**

*I found a nice tea set at*
**makerworld.com/en/models/434978-tea-set**

*Search for "miniature tea set STL" and you'll find a few to choose from. Some are very clunky but keep looking or make your own.*

*The door knobs on the cabinet are just cubes with a pyramid and sphere on top.*

# Belfast Sink

The sink will be Belfast style and the tap is an old, fiddly, Victorian thing. Here's my planning diagram. The double green line are markers for the centre and inner curve of the corner.

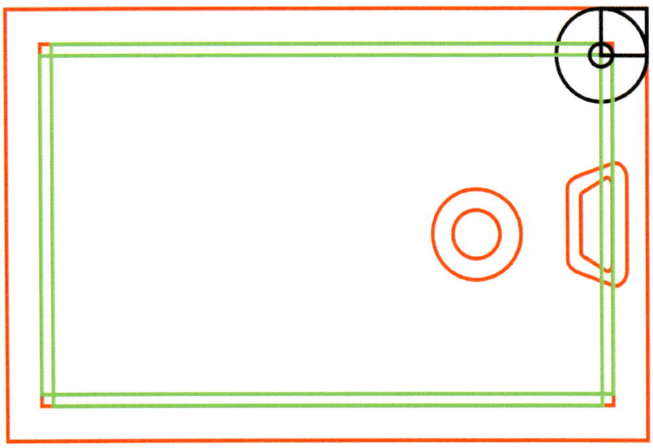

The draining board we just made slopes in towards the sink from sides and back and has a plateau for the tap.

I have taken this design from an old photo where it doesn't overlap into the sink for easy draining, but Angie insists it should, so you may want to shrink the height of the sink and extend the draining board a little for your project.

Start with a cube and make it 3 mm wide by 29 mm and 21 mm high to make the side wall (shrink this for a shallower sink if you wish).

Make a "round roof" piece of 3 mm wide by 29 mm and 2 mm high.

Align it to the top of the side wall then lift it 2 mm to sit on top and curve it.

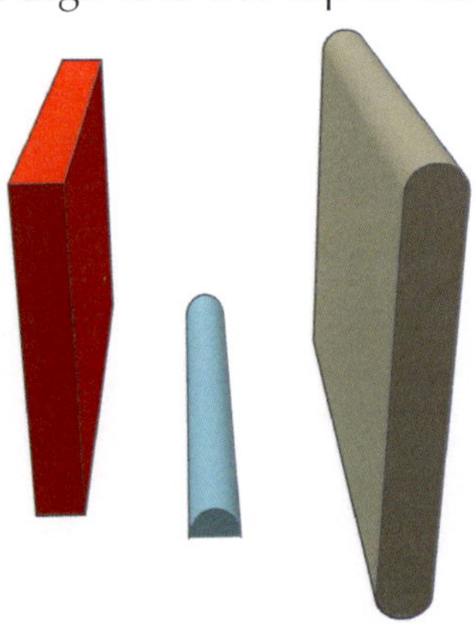

Lift both pieces 2 mm.

Make another round roof and rotate it 180 degrees over. Align it to the bottom of the side wall and bundle the three pieces.

Make a copy of the wall and move it to the side.

Make a copy and rotate it 90 degrees. Now lengthen it to 48.5 mm wide. Copy it.

We now have 4 walls to the sink and need to make a smooth, rounded, corner.

So far everything would have been a lot easier in software (e.g. Blender) with a bevel function (rounding off sharp edges to objects). However once we come to the corners it is easier to use Tinkercad.

After a lot of experimenting I think the fastest and easiest way is to cut a torus plus cylinder into quarters.

Make a cylinder of 8 mm and 23 mm high.

Make a hole cylinder of 2 mm and 23 mm high.

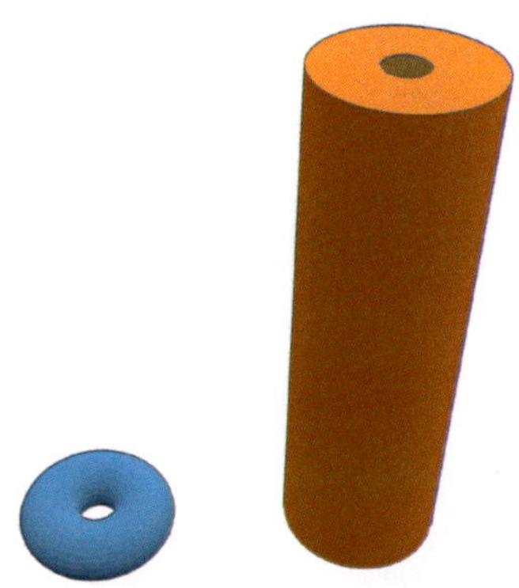

Make a torus of radius 8 and tube 5. Make it 8 mm and 4 mm high.

Align the torus to the top of the cylinder then move it up 2 mm.

Union the 3 objects.

Make a hole cube of 4 mm wide by 8 mm and 25 mm high.

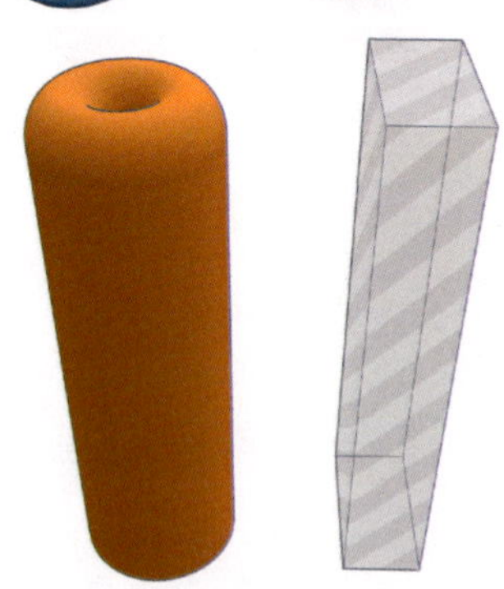

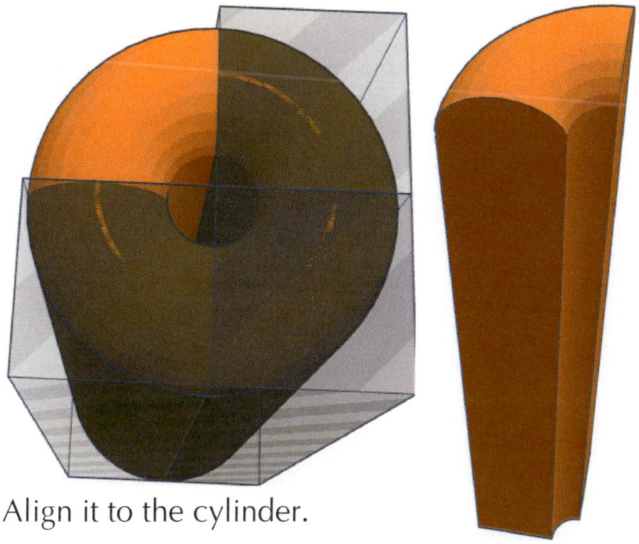

Align it to the cylinder.

Copy the hole and rotate it 90 degrees. Align it.

We are now ready to select the 3 objects and Union them to make a corner piece.

Unlike the walls I am letting this one go to the baseline at the corners so it looks like feet.

Copy and rotate 90 degrees around the base. Repeat for two more corners.

For the base make a cube 56.5 mm wide by 37 mm and 1.5 mm high.

You can now align all the corner pieces to the appropriate corner, and all the side pieces into the centre of the base cube. Bundle everything except the base.

Shrink the base cube to 53.5 mm by 34 mm and centre it.

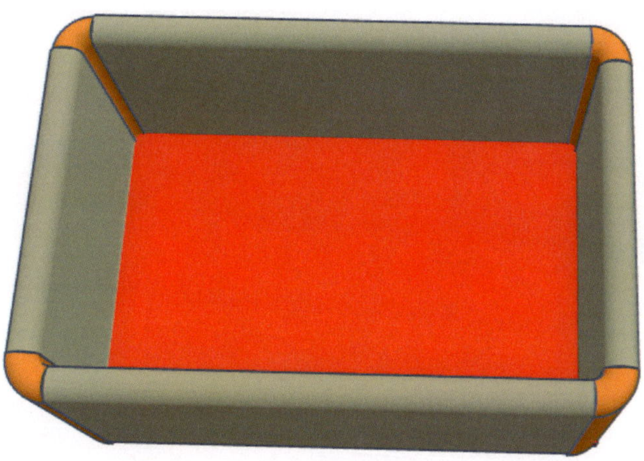

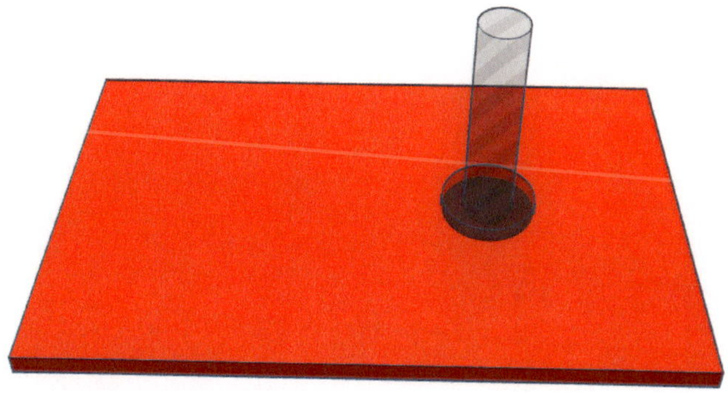

To make a plughole we need to put a hole cylinders in at the centre and as far towards the right as looks good to you.

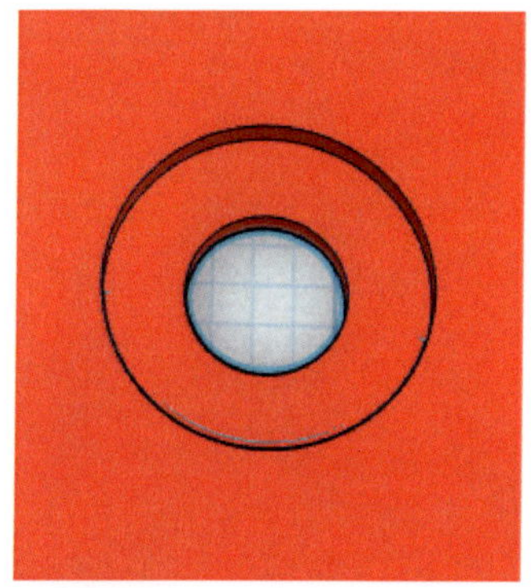

Make a hole cylinder of 8 mm and 0.75 mm high and align it to the top of the base.

Make a hole cylinder of 4 mm and centre it to the wider cylinder.

Select the base and holes and union them.

Finally we are going to make an angled overflow for the side with the plughole.

I found this easiest in Inkscape as I could make primitives roughly right in Tinkercad but not easily smooth their edges.

Make a box of 5.25 mm wide by 10.5 mm, colour it black and click the node tool.

Using the round node at top right pull it around a little to curve it.

Now click Path -> Object to Path.

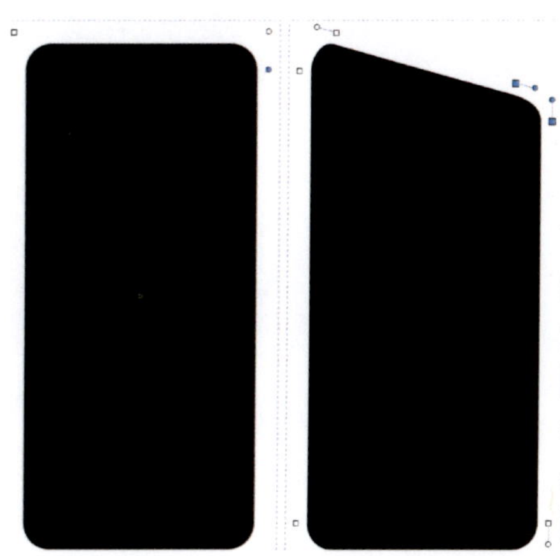

Make a copy and colour it white.

Using the Transform box you can Scale it, e.g. height 70% width 80% or whatever looks good.

When you are happy click Path -> Difference and save it as sink overflow.

*printed sink*

Import this in to Tinkercad and make it 20 mm high.

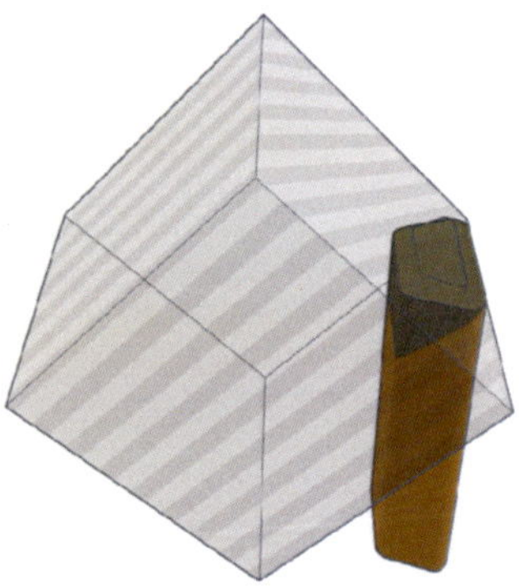

Make a hole cube and rotate it slightly - approx 30 degrees - and align it to the top of the overflow. Move it up until it is taking a slice from the top.

Union them.

You can now align this into the sink at the centre right.

Select and look around the sink from all angles to make sure it is perfectly aligned before printing.

# Pots, kettles and taps and table & chair

## This project teaches you how to:

- *Layer simple shapes to create complex results*
- *Advanced combinations of primitives*
- *Table & chair recaps what we have learned so far*

# Cooking pot

If you saw the pots and vases in the first book this pot will be familiar but a little extra. The kettle to follow will be more detailed still. The ultimate example of this building with primitives technique will be the antique tap in the next project.

So to start the pot we are going to take a sphere and cut off the bottom and top to get a nice curved section.

Make a sphere of 40 mmm.

Make a hole cylinder of 34 mm and 7.5 mm high.

Align it to the top of the sphere, copy it and align the copy to the bottom of the sphere. Union all 3.

Make a hole cylinder of 37 mm and centre it in all axes to the sphere. Union them.

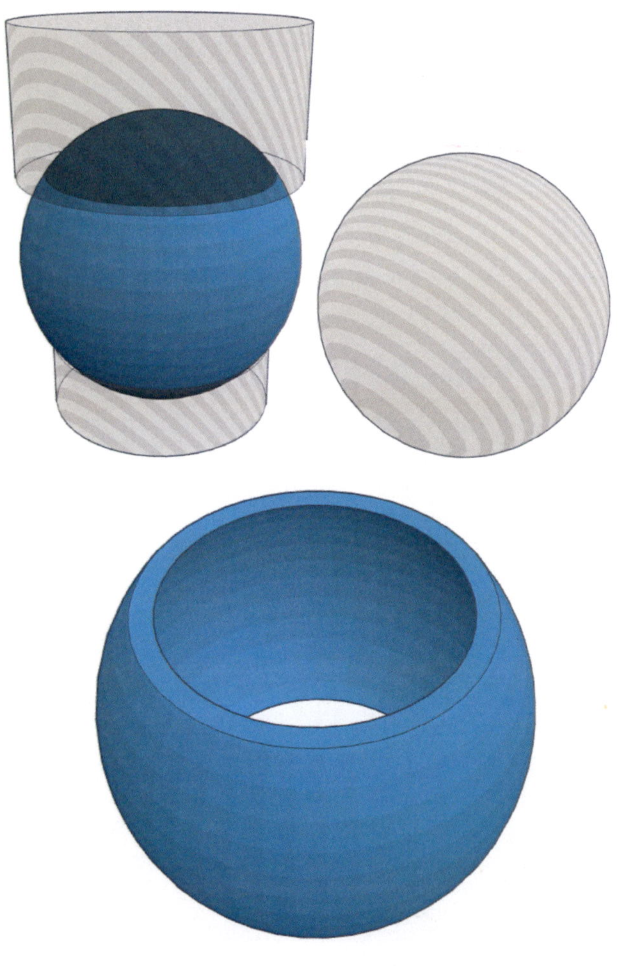

We now have a pot and need to add a base and a lip. Move it down to the baseline (0.00) with the black movement cone on top.

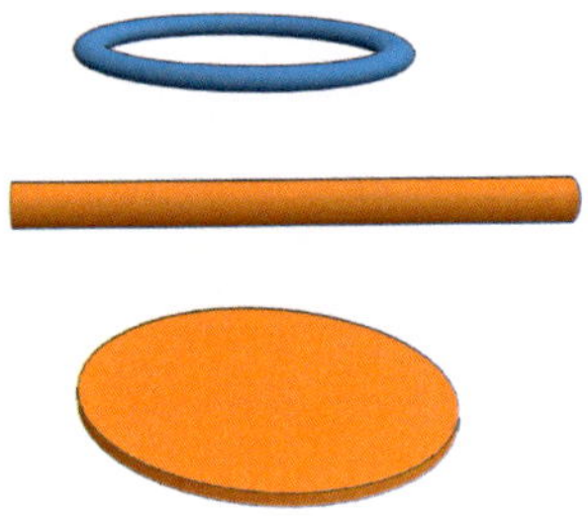

Make a cylinder of 32 mm and 1mm high. Centre it to the pot.

Make a torus and give it radius 7.5 and tube

0.75. Make it 32.2 mm and 1.9 mm high. Align it to the top of the pot.

Raise it 1.5 mm with the black cone.

Now to add a handle. Make a cylinder of 3 mm by 3 mm high and rotate it 90 degrees. Raise it to a little over halfway up the pot. Align it to the centre of the pot and move it in until you can see the end is embedded in the pot but not poking through to the inside. Copy it to the other side.

Make a torus with radius 7.5 and tube 0.4. This is quite thin and will need to be printed

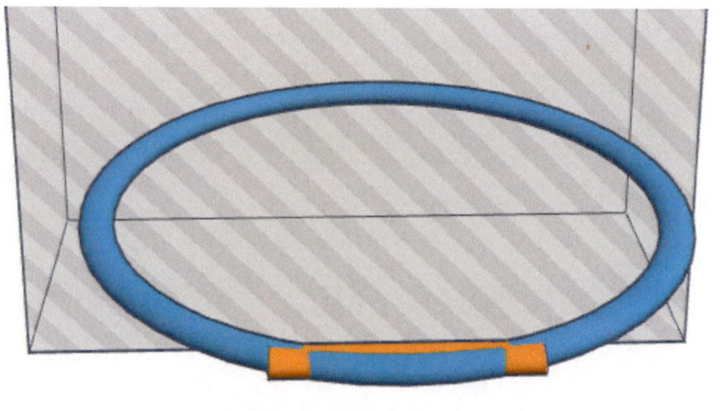

with a brim. Make it 43.75 mm by 39.73 mm and 1.3 mm high.

Make a hole cube of 45 mm by 20.5 mm and align it centre, back to the torus.

Make a cylinder of 2.5 mm by 1.3 mm and 15 mm high. Rotate it 90 degrees so that it is 1.3 mm high.

Align it to the top, centre of the handle torus. Move it inside a little until it looks like a handgrip.

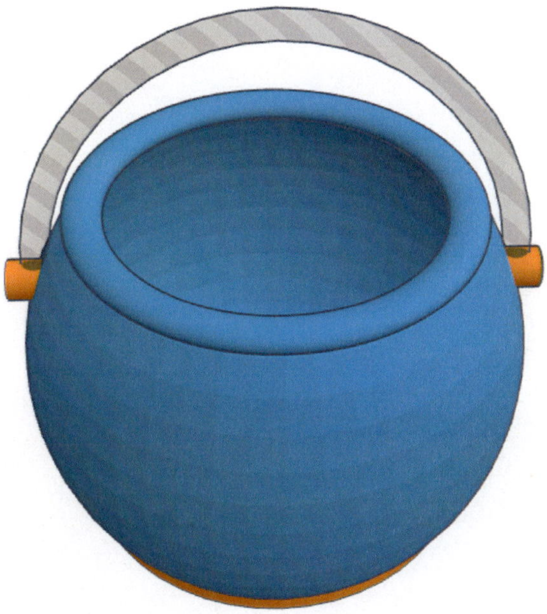

Union these up and make a copy. Rotate the copy 90 degrees upright and make it a hole.

Align it centre to the pot and raise it until it pokes through the 3 mm cylinders. Union the pot, base, lip and handle pieces. If you wanted you could make slightly bigger holes here to make it easier to fit later, I did find this a bit tight.

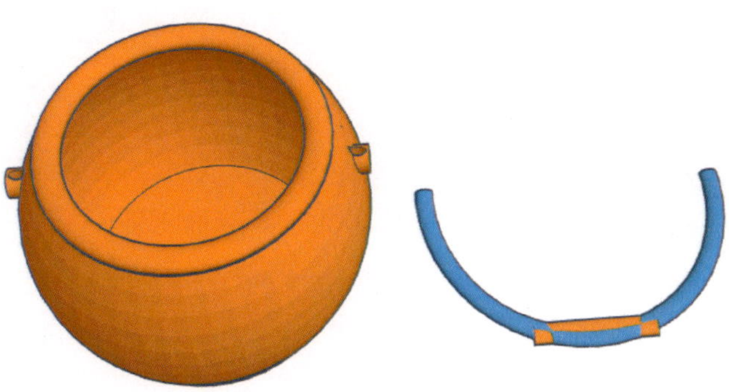

At this point you can export and print the pot. Print the handle as a different job with Brim enabled.

## Kettle

Now to add a little more complexity. An antique kettle with a spout, handle and lid.

The body will be a merge of two shapes - a cut down cone and a semi-sphere.

The spout will be a torus chopped up and spun back around on itself to give the characteristic shape.

The handle will be another torus cut in half. We will add little details too along the way.

So starting with the base we need a cone of 30 mm and 60 mm high.

Make a hole cube of 30 mm wide by 30 mm and 45 mm high, align it to top, centre of the cone and union them.

Rotate the result over 180 degrees so it slopes down slightly to the base.

Make a semi-sphere of 30 mm and 15 mm high and move it up off the baseline by 15 mm.

Centre the two pieces together and bundle.

Make a hole cylinder of 15 mm and 30 mm high, and raise it 2 mm off the baseline.

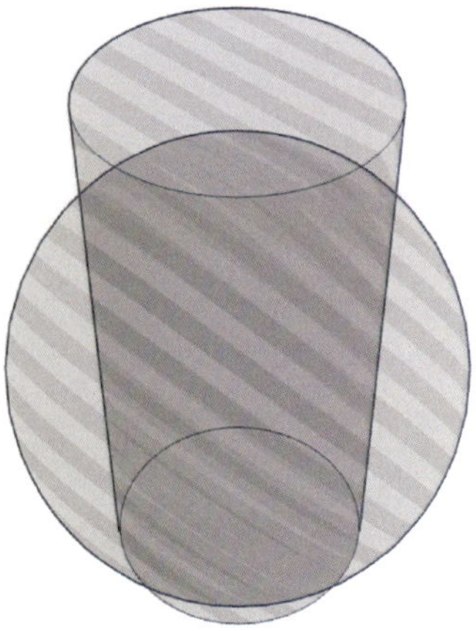

Make a hole sphere of 25 mm, and raise it 2 mm off the baseline.

Centre the bundle and the two holes together. Union them.

For the spout make a torus of radius 7.5 and tube 2.5. Make it 20 mm and 5 mm high, then rotate it on end 90 degrees.

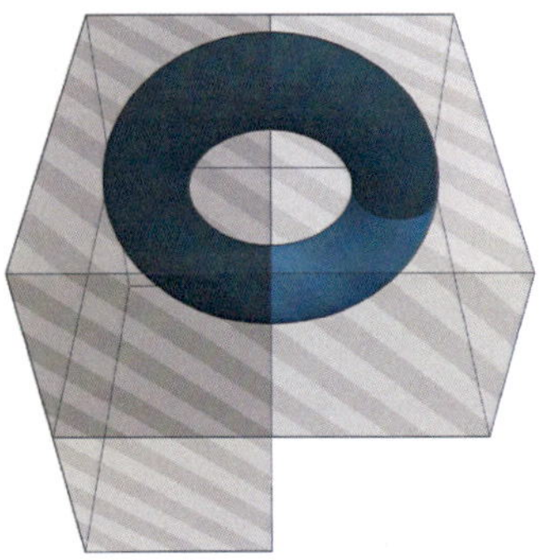

Make a hole cube of 10 mm by 20 mm and 20 mm high and align it left to the torus.

Copy the hole, rotate 90 degrees and align it top and centre.

Union the three pieces.

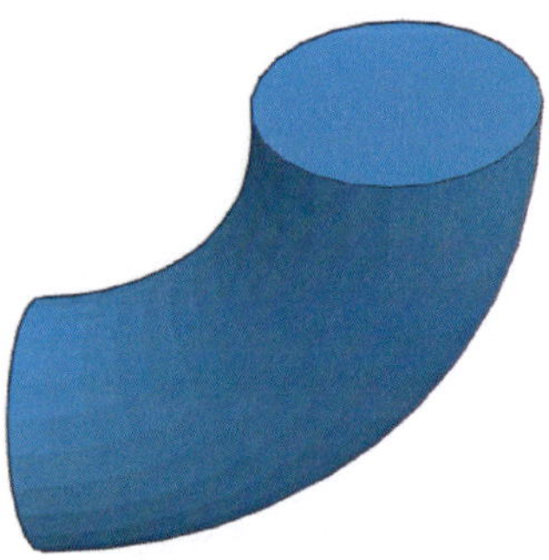

Copy this quarter tube and rotate it 180 degrees, move the piece up 10 mm and arrange it with the arrow keys to fit perfectly on the first piece. Bundle them.

Make a hole cylinder of 3 mm and 2 mm high. Rotate it 90 degrees and move it up to reach near the top of the spout. Align them centre with the hole at the right of the spout.

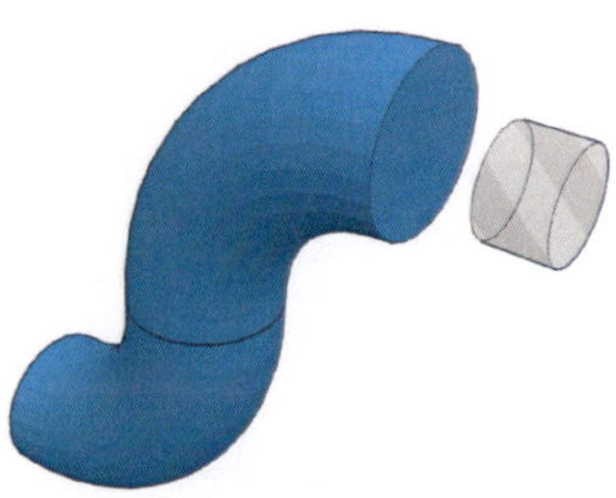

Make a wedge and turn it into a hole. Rotate it 90 degrees and move it up to meet the pout. Use the arrow keys to push it in just enough to leave a V shape. Union the spout and holes.

Finally on the spout we want to ramp it into the body a little so make a cone of 8 mm and 8 mm high. Rotate it 90 degrees pointing to the back of the spout and centre it.

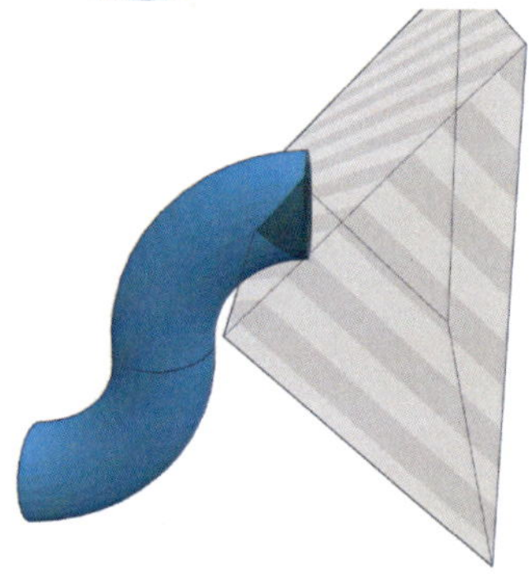

Move it in with the keys until it looks right - leaving 2-3 mm visible. Bundle them.

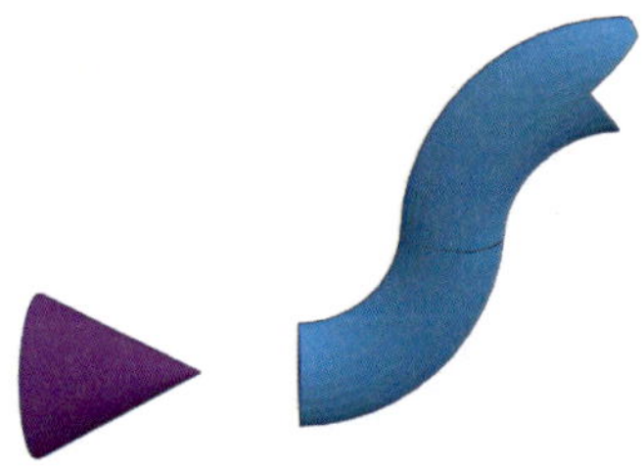

That leaves handles and a lid to make. I have put a couple of small wedges into the body for the handle to go into, but you can choose anything you like for this part. Maybe cylinders would look better or a rounded rectangle. It is just to look like a plate that the handle is welded onto.

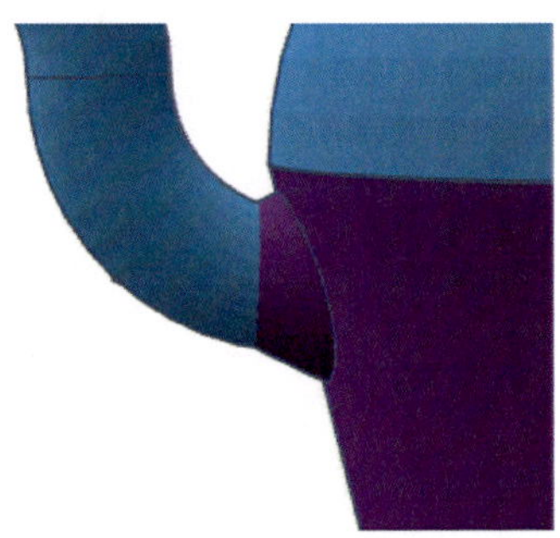

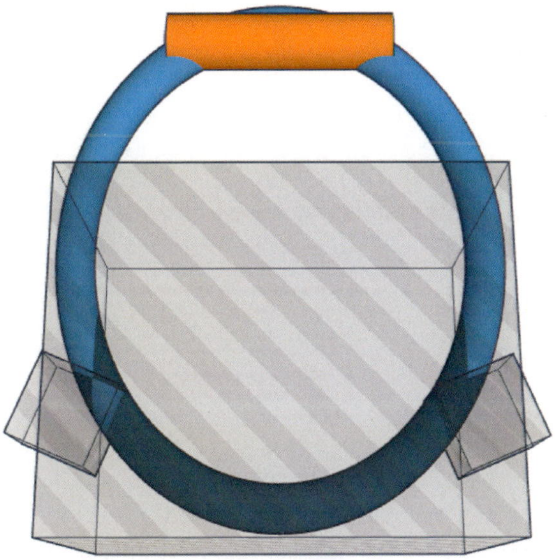

You now have all the pieces to export and print, remember to put Brim on the handle.

The handle itself is a torus of radius 9 and tube 0.8, make it 28.5 mm by 33.75 mm and 2.5 mm high.

Make a hole cube of 23 mm by 17 mm and align it centre, bottom to the torus. Union them.

At this point you can also trim the ends flush with holes if you wish.

To make a handgrip make a cylinder of 3.5 mm by 2.5 mm and 15 mm high. Rotate it 90 degrees with the narrow part upwards. Align it to top, centre of the torus then move it down a little.

Like with the pot handle we can union this and make a copy as a hole to cut out where the wedge plates are, or we can just "drill" slightly large holes there with cylinders.

The lid is a simple cone of 15.5 mm and 20 mm high, with a hole cube of 18 mm high aligned to the top.

Union them and flip the lid over 180 degrees so that it slopes down slightly.

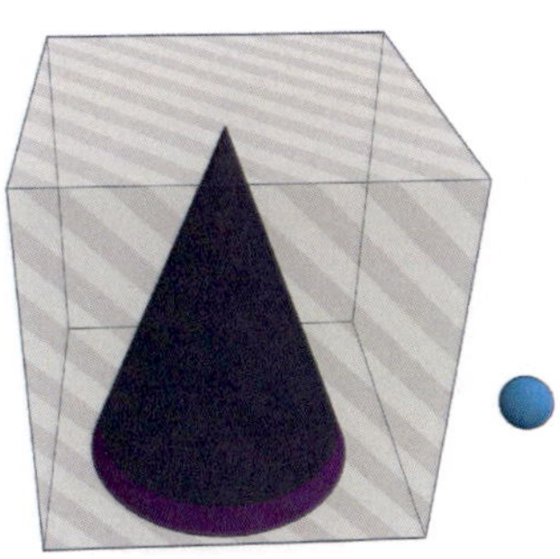

Add a sphere of 3 mm raised by 1 mm off the baseline. Centre and union.

# Tap

We are going to do a fast run-through on assembling a complex shape from simple shapes.  It is possible you might need to read through this a couple of times during assembly. Just remember at any time you can change pieces, re-align things or just press undo if you aren't sure what happened.

> *If you are of a certain age, back in the 1980s you may have typed in lots of code from magazines into personal computers. This will feel like that except at each stage you can see a result - you don't have to wait until the end only to realise you made a mistake!*

## Tap base

For the tap base make a cylinder of cylinder 6 mm by 0.5 mm high.

Make a cone 6 mm and 3 mm high.

Make a cylinder of 3.5 mm and 4.7 mm high for the body.

Make a cone of 3.5 mm and 5 mm high.

Make a cone of 2.6 mm by 5 mm and 4 mm high, rotated 90 degrees to point left-to-right and align the vertical centre with the cylinder body.

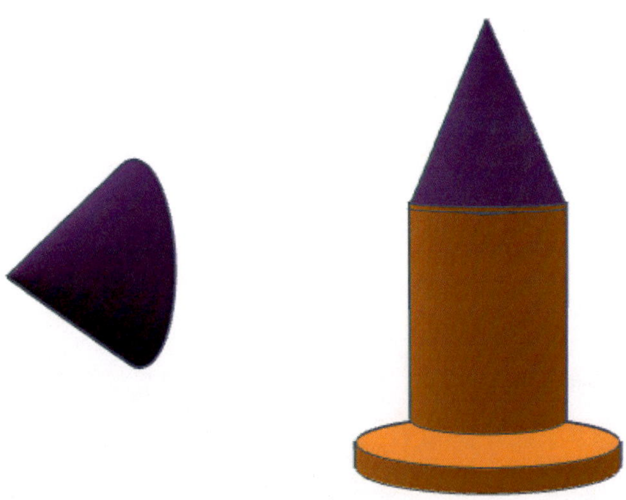

## Tap handle

For the tap handle make a cylinder of 2 mm and 6 mm high.

Make a cylinder of 1.5 mm and 8 mm high, rotate 90 degrees to horizontal so it runs left-to-right. Copy it and rotate 90 degrees to form a cross.

Make a sphere of 3 mm for the centre of the cross.

Make spheres of 2 mm and 3 mm high on end of each cross cylinder.

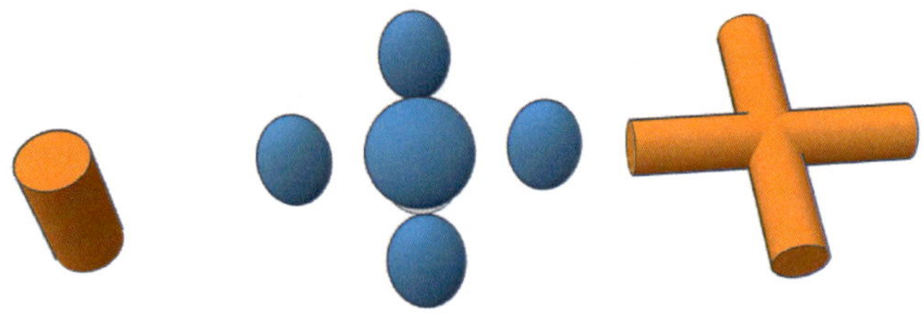

Make a cylinder of 2.5 mm by 3 mm and 23 mm high.  Rotate it 90 to horizontal to run left-to-right, align it centre and left to tap base. This make a crossbar between the two taps and also holds the centre spout.

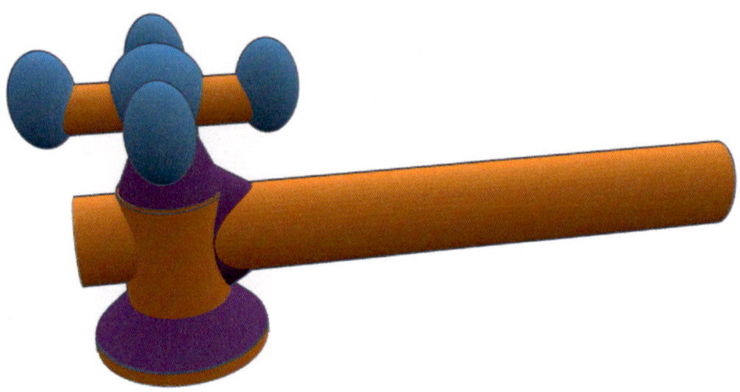

Position and align everything except the cross bar to make a tap body and handle. Bundle it. Copy it for the second tap.

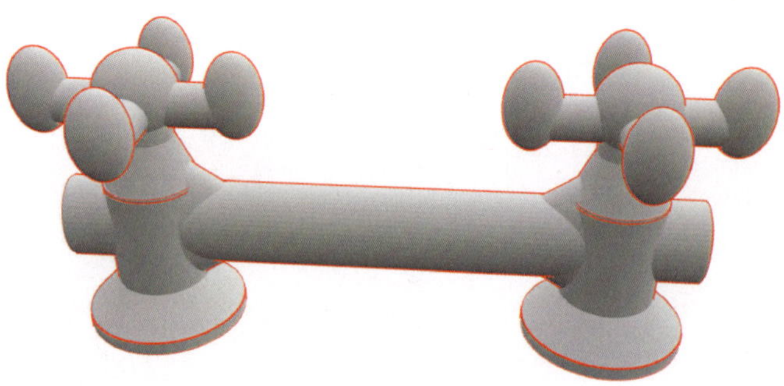

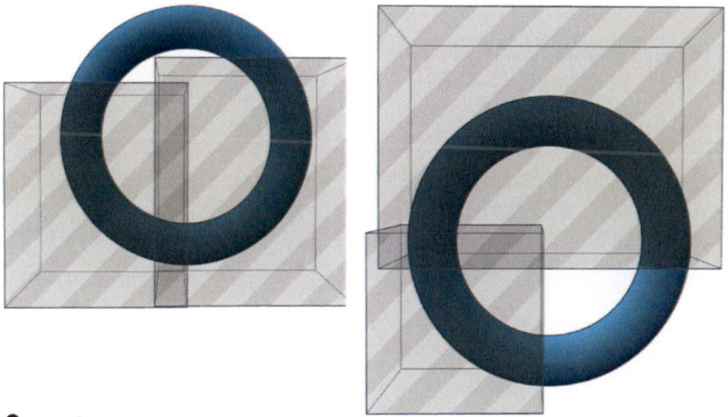

## Spout

For the spout end-piece make a cylinder 2.5 mm and 3 mm high.

Make a cylinder 3.5 mm and 1 mm high. Align both to centre, top.

Make a hole cylinder 1.75 mm and 2.5 mm high and align to centre, bottom. Union the three cylinders.

Make a cylinder 2.5 mm by 9 mm high, rotated 90 to the horizontal to run left-to-right.

Make a torus radius 15 tube 2.8 size 15 mm and 2.8 mm high

Make two hole cubes and put one at the left around 2/3rds up, one at the right about 4/5th up. This doesn't have to be exact.

Copy the torus and move one away.

Union the torus and the two hole cubes.

Take the copy of the torus and cut it roughly into a quarter using hole cubes. Union.

Align the pieces to make a spout.

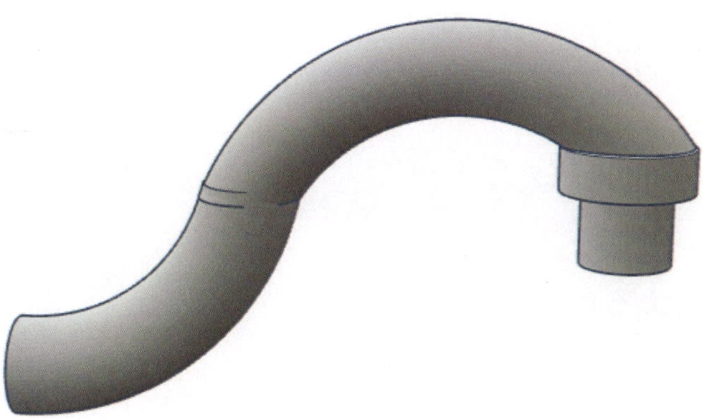

## Centre tube

For the centre tube make a sphere of 6 mm by 4 mm and 3 mm high.

Make a cone of 4 mm and 6.8 mm high.

Make a cylinder of 3.75 mm and 2 mm high

Make a cylinder of 3 mm and 2 mm high.

Make a sphere of 3 mm.

Make a cylinder of 3 mm and 6.75 mm high.

Make a torus into a hole with radius 7.5, tube 2.5, make it 5 mm and 2 mm high.

Make a cylinder of 3 mm and 1 mm high.

Make a sphere of 4 mm.

Make a cylinder of 4 mm and 2 mm high, rotate it 90 degrees to present a flat face

Make a hole cylinder of 2.5 mm and 2 mm high, rotate it 90 degrees as before, centre it to the 4 mm cylinder.

Make a cone of 2 mm and 2.5 mm high.

Make a sphere of 2 mm.

At the back make a cone of 2 mm and 3 mm high, rotated 90 degrees to point away from the tube.

Make a sphere of 1.5 mm aligned to the point of the cone.

Centre all the pieces down the L-R middle and position them to look good using the image on the next page. Union them.

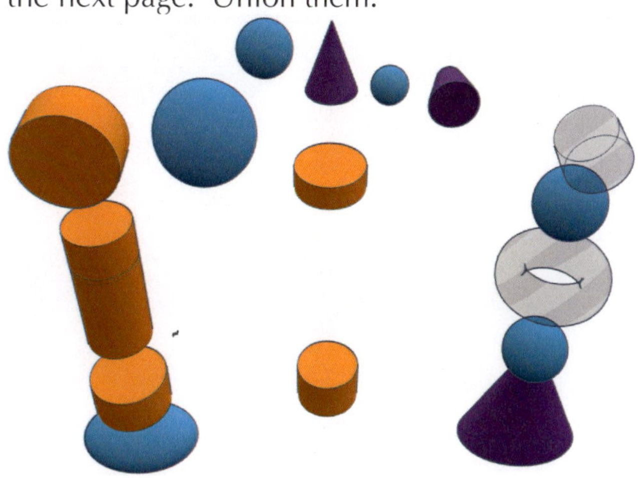

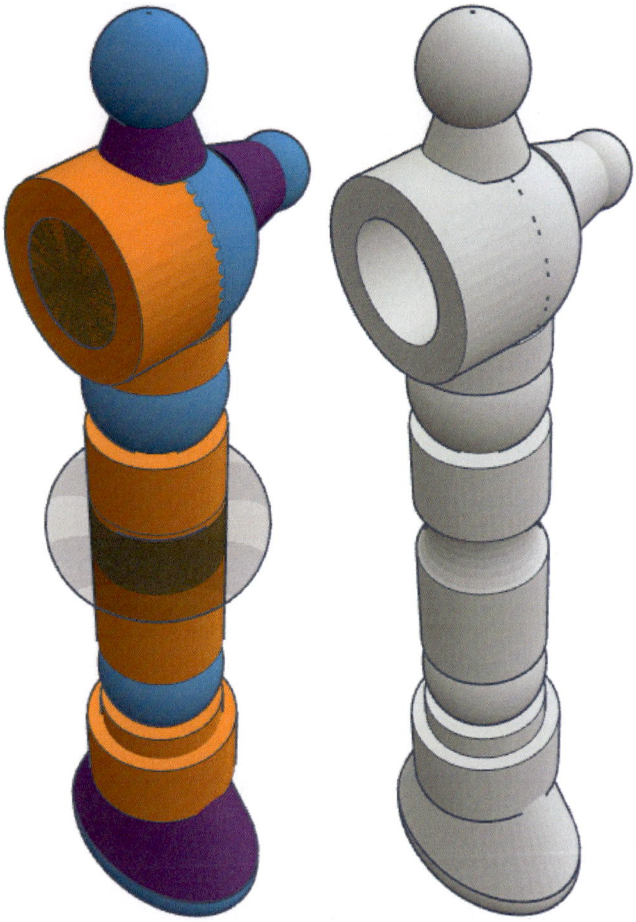

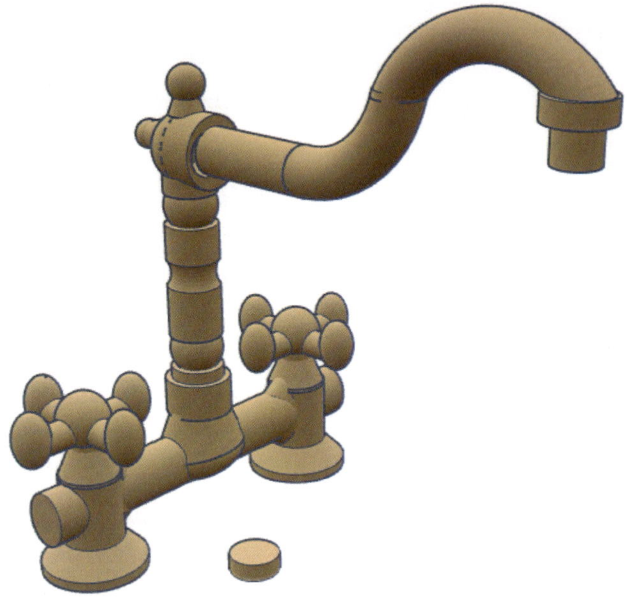

Now assemble the spout, tap bodies and centre column to make a complete tap.

# Table

This one is a reprise of what you have already learned - I am just going to show the parts, rotation and sizes, you already know how to assemble them tidy and finished. This is a grand kitchen table to hold 6-8 people. You can scale it down a little to suit by narrowing it as you build the supports and table top.

## Table leg

First make a cube of 14 mm all around.

Make a tube of radius 10, wall thickness 2.5 and 28 mm wide, turn it into a hole.

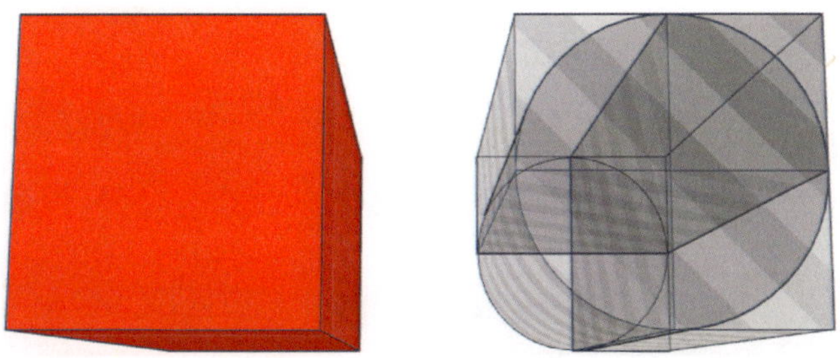

Move it so it takes a small curve out of one corner of the cube then Union them.

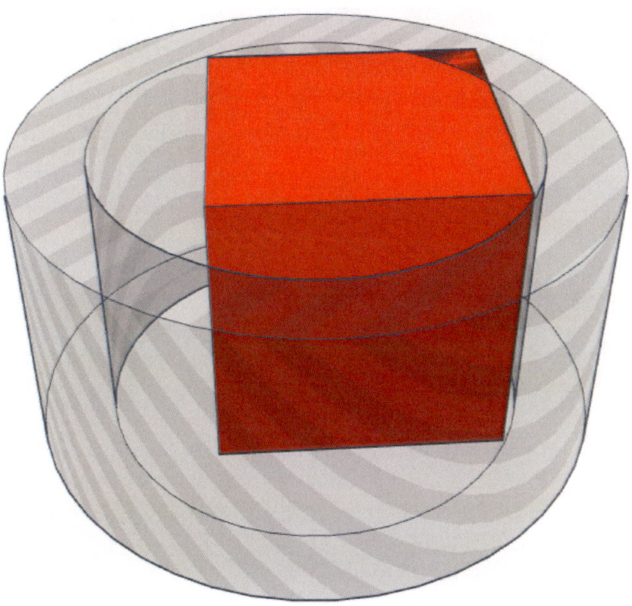

Make a cylinder 6 mm and 46 mm high.

Make a cylinder 8 mm and 28 mm high.

Make a torus with radius 7.5 tube 2.5 and make it 10 mm and 5.25 mm high. Copy it and move it higher up.

Copy it and make the copy 8 mm and 2.5 mm high. Copy the copy twice more.

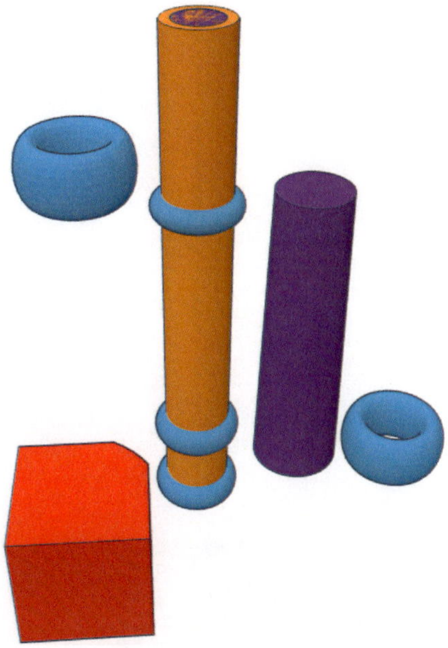

Align all the pieces to the centre and adjust them up and down to match the diagram.

Union them. We now have a table leg (but upside down for now).

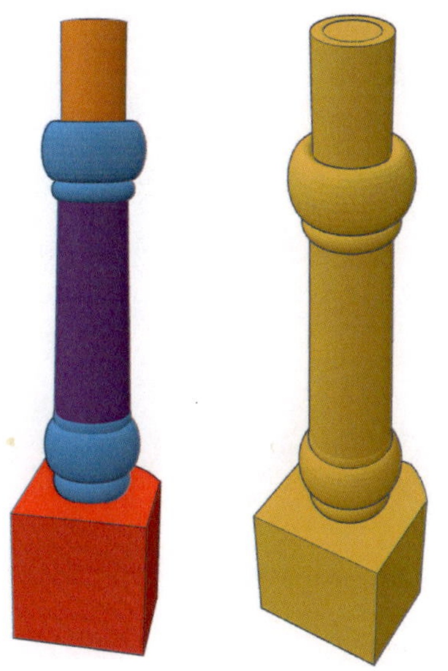

## Table under-support

To make supports make a cube of 80 mm wide by 4 mm and 14 mm high. Copy it and move it away.

Copy it again and make it 186 mm wide, then rotate 90 degrees. Copy it.

Align the pieces to the four corners to make a frame.

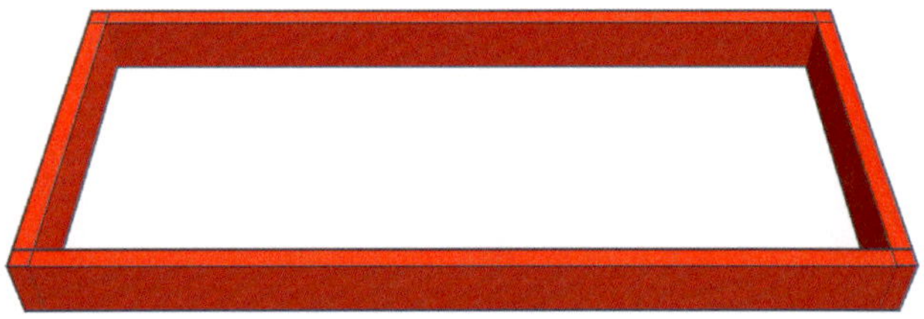

Align a leg so that it's curved corner faces outwards and move it a little outside the frame as shown - not flush.

Copy and rotate to each corner and union the legs and frame.

## Table planks incl scoring

Make a cube of 212 mm wide by 106 mm and 4 mm high for a table top.

Make a hole cube 212 mm by 1 mm and 1 mm high. Copy it twice and space them evenly across the table top. Align to the top.

Make a hole cube of 9 mm wide by 1 mm and 4 mm high. Copy it and align to each end of the three hole cubes we just placed.

These will score what looks like planking into the surface.

Finally to round off the edges of the table Make a cube of 8 mm.

Make a hole cylinder of 8mm.

Make a hole cube of 8 mm by 4 mm, copy and rotate it. Align all to the corner of the cylinder. Union them.

This should leave a curved piece - turn it into a hole, align it to each corner of the table top and rotate 90 degrees to take it to the next.

Union the table top and hole pieces.

You can print the table top and legs in different coloured wood, I went for dark on both initially but Angie insisted a kitchen table would have a lighter coloured (pine) top.

# Chair

The chair leg is easy, we can the shrink one we already made for the table.

Copy a leg, shrink it to 10.5 mm wide and 40 mm high.

Angle (tilt) it by 7 degrees, front to back and left to right (just as we did in the stool project) towards the rounded corner.

Make 3 copies and rotate them so the rounded corner is to the outside on each one.

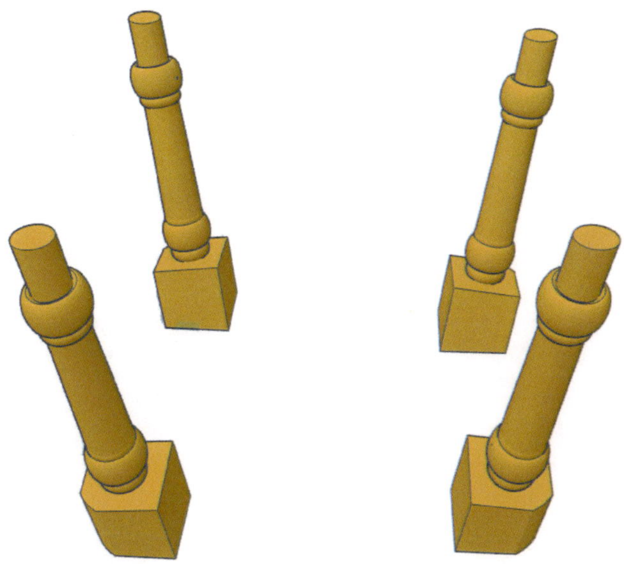

## Chair seat

For the seat I took circles, merged them in Inkscape and curved out the result until I was happy - narrower at the back than front and about 40 mm square overall, 4 mm high.

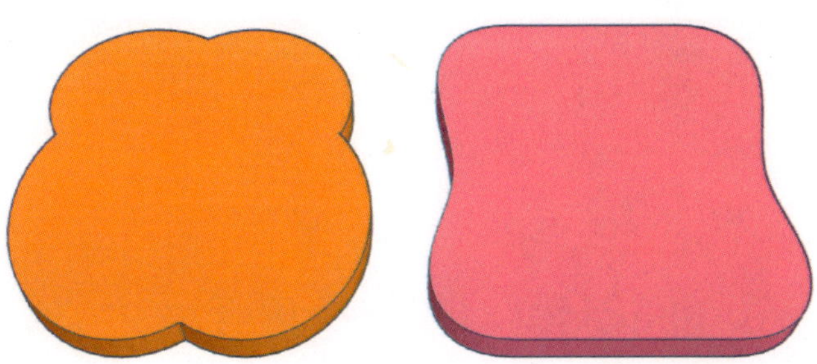

## Chair seat grooves

The grooves were flattened spheres and cylinders angled 7 degrees out towards the edge, mirrored and joined together at the surface of the chair to cut a groove. This needs a little sanding on a filament printer.

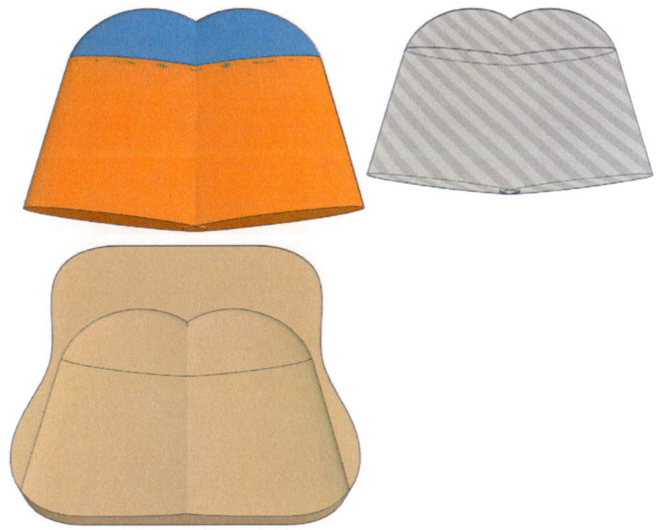

## Chair back

The back is your choice but I went for a hoop with 3 support posts.

The posts are cones chopped to length and then put back to back to be thicker in the middle. You can use the stool legs from the first project and shrink them to 40 mm high. Copy and rotate 180 degrees, Align and Union.

The two outer ones are angled 7 degrees. I cut them to size using hole cubes after angling them.

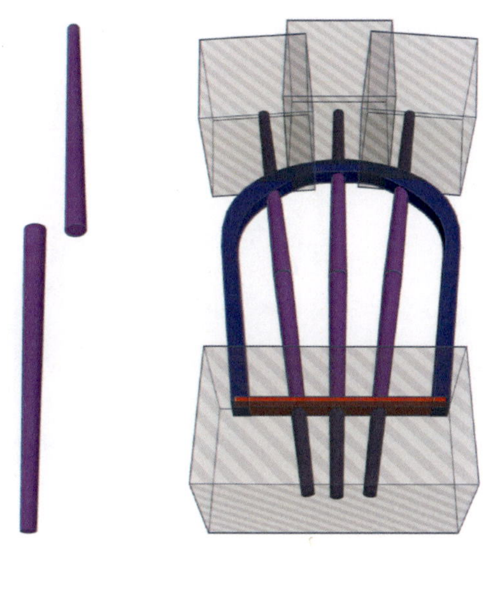

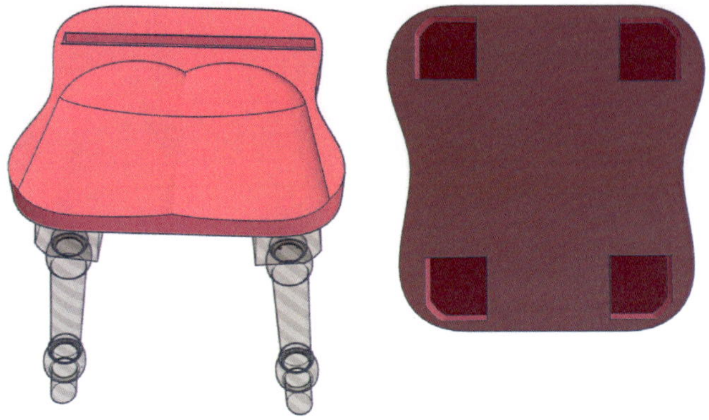

I tidied up the legs using a cube above and below them to make them perfectly level. I then copied this and cut it out below the seat to leave them nice firm holes to click into.

The hoop at the back was done in Inkscape for ease and is 44 mm wide by 52 mm and 3 mm high.

Finally I stretched a cube across the bottom to make a bar joining them all together.

## Assembly

I copied and angled the finished back at 7 degrees out away from the chair and used it as a hole to cut a groove into the seat for later glueing.

# Tiles and trim

## This project teaches you how to:

- *Combine patterns into multicolour prints*
- *Use filler for extra colours on single colour printers*

# Tiles and trim

In this section I am going to cover all the little details that make a scene look good. Coving, panelling, trim, tiles and ceiling roses.

If you have done the introductory projects you should have the ability to achieve any of these, as we have covered all the skills needed there.

## Encaustic tiles

You can print multicolour tiles easily if you have a four colour unit, but you can also do it with a single colour machine and filler.

If the project is for personal use you can download any design you like from free vector art sites on the internet and scale it to the size of tile you are making. If you want to sell the design it might be best to make the images yourself. Even if it is an historic design like the Fleur De Lys the vector you download might be copyright, so I designed my own.

I have chosen 4 tile designs. A couple of geometric styles made of lines, a floral pattern and a quarter tile Fleur De Lys.

### Tile 1

I have done the first geometric design in Inkscape here but the geometrics are just as easy in Tinkercad. The first tile is simple rotations of boxes and rectangles at 45 degrees arranged inside a square. The final tile is likely to be one inch or 25 mm so it leaves room for plenty of detail.

The stars dotted everywhere are simply boxes rotated 45 degrees and placed at each intersection. They can be a different colour to the main tile too, e.g. red.

Draw the inch box and make a rectangle twice as long but 10% deep, e.g. 2.5 mm wide by 50 mm. Rotate it 45 degrees.

Copy the box and rotate it 90 degrees to make a cross.

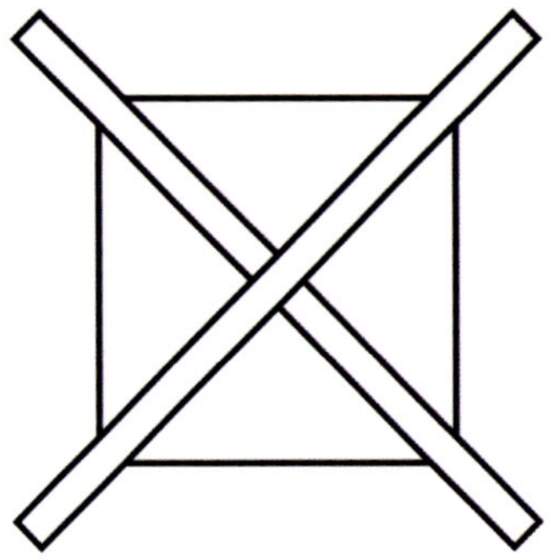

Copy the cross and move it until the intersection is just outside the box at the top.

Copy the cross and move it until the intersection is just outside the box at the bottom.

Copy the inch box

Select all the cross pieces and Path -> Union them, then Path -> Intersection.

Now make a box of 2.5 mm, copy and rotate 45 degrees, union the two. Now you have a star to place at each intersection.

At this point I shrank the whole design a little to leave a 12 mm border thinking that it might be useful when making multiples join. In which case put the final star only a quarter present at the corners and half at the edges.

Once you are happy with the design you need to union all the pieces for each colour of print. If you can only print a single colour use filler for the second colour.

Export the base of the tile and the pieces and bring them into Tinkercad.

Make an inch square cube (shown in red). Depending how thick you want the result set the height - I suggest 3 mm high for floor tiles.

Make the first colour pattern 2 mm high (shown in white), copy it and make it into a hole. Align this to the centre, top of the tile and Union to cut it out (shown in grey).

At this point if you are going to fill it manually you are ready and can print the tile.

If you intend to print it multi colour export it.

Take the colour pattern and add a 3 mm cylinder at 0.1 mm high.

Align the pattern top, centre and bundle it with the small cylinder. This fixes the height for when you put it into the slicer and is important. Export the pattern bundle.

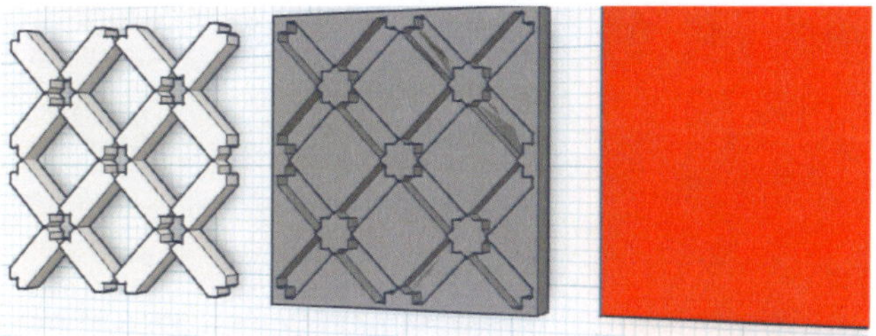

In the slicer import the two (or more) pieces, check they are perfectly aligned and assign them colours. They are ready to print.

## Manual infill

If you have a single colour printer and want to fill in complex designs we have tried a couple of options successfully. The most important thing when printing is to use the Ironing setting on the slicer so the top face of the print is smooth and not textured.

For the white geometric tile we used stucco filler from a small pot - ideal for craft uses.

Simply apply with a spatula and leave to dry. If it dries back too much apply another coat. Once you are happy scrape it back to level with a blade (single sided ones are easiest to use). Progressively sand it with nail files right down to buff and smooth. If the surface is still not shiny you can apply a little varnish.

The metallics were UV nail gel and metallic powders mixed then applied with a spatula. These need to be cured but a sunny day on the windowsill is plenty.

They don't take a lot of sanding down, just work all the way down to fine then buff then shine. Again if you want a shinier result you can varnish.

## Tile 2

This is the other geometric design and is three tiles wide making a pattern that goes across all three.

Again using simple rectangles of approximately 10% of the tile height. I have laid them out to make patterns and boxes. The only rule is that the spacing must always be the same.

At the start and end of the tile I have left a tiny border for using manual infill to stop it leaking out. It isn't necessary if printing in two colours.

Follow the method for tile 1.

## Tile 3

This tile is easiest in Inkscape or something with sketching and editing abilities, it is not suited to Tinkercad.

Tile 3 is a floral design based on an old tile I saw. The petal parts are largely the upper parts of the Fleur De Lys I made for tile 4. I rotated and copied them round 90 degrees for all the sides.

There is a semi-circle design overlapping above the petals and two inner circles to complete it.

Then I added a rounded arrow for the corners.

## Tile 4

This is a simple Fleur De Lys at 45 degrees on a quarter sized tile -. half inch / 12 mm.

You can often download this design for personal use.

*3D printed tiles, multicolour*

### Notes on the finished pictures

The finished shots show some trouble with the metallic filament in the multicolour prints, it was probably a little damp and bubbly at fine resolution. There is also a problem with the terracotta and green version of the multicolour too, probably for the same reason - the filaments had been on the machine a few weeks uncovered. The problem with miniatures can be how little you use!

The manual infill of UV gel and metallic powders had a nice aged look to it and I probably prefer it, but I wouldn't want to fill a large area of them.

## Doorstep checkerboard

This one is easest in Tinkercad, a simple 11 x 12 grid of squares. I chose to go with the equivalent of 70mm (3") tiles for this and made a square of 5.83 mm, then rotated the square 45 degree. Duplicate a line of 11. using the arrow keys to position them, then group them.

Duplicate the line down with the arrow keys to make 12 and group them.

Now make a border square of 100 mm by 110 mm (the size of the doorstep), and cut into it (Path -> Difference) with another square exactly where the tile edges touch. Select all of this and export to Tinkercad.

Make the tiles the same height as the recess we cut in the doorstep - 1.3 mm.

### If you are only printing single colour:

add a base layer of 100 mm by 110 mm and 0.3 mm high to hold the infill.

### If you are printing multicolour:

Copy the tiles and make it into a hole.

Make a cube of 100 mm by 110 mm and 1.3 mm high, align and union it with the hole to make the white infill.

Export the tiles to STL and the infill to STL. In the slicer align them perfectly and assign colours.

Finally as I decided to cut into the main step for fitting the columns I have done the same to the checker pattern.

## Porch wall tiles

This is quite a complex decorative piece that I only just got finished in time for the book. This is definitely a multicolour print. You can simplify it to plain tiles and a surround which would be perfectly normal and achievable with a single colour printer.

First decide how many panels you have and make tile sections to add the design to. We went for three panels - mirrored outer flowers and a central one.

Angie drew the design and I worked on a picture of it in Inkscape to simplify it and make it fit the 3 tiles I had assigned to it.

This makes a 3 panel set of lilies which I will reproduce in 4 colours: terracotta for the vase and petals, green for the leaf and stem, white for the tiles and brown for the surround. Metallic bronze for the surround would also be good.

The surround is a dado piece at the top and skirting at the bottom, with plain pieces in-between to space out the white tiles.

There is a lot of organisational work for a print like this. Collect all the terracotta and green coloured pieces into a bundle. Copy them, make into holes and cut from the backing tiles.

Make a thin 0.5 mm high backing plate to hold all the pieces together and raise everything else on top of this. You may want it brown under the surround so this will be another cut-out task.

Export each colour group and assign a filament to it in the slicer.

Finally print on a high quality setting to make the most of crisp joins between pieces.

## Panelling with bead, skirting and dado rail

Simple beaded wood panelling for the walls with a dado rail and skirting board.

We can re-use the simple architrave from above the doorway as a dado rail and amend it with a 3 mm square cut out, to match the height of the panelling.

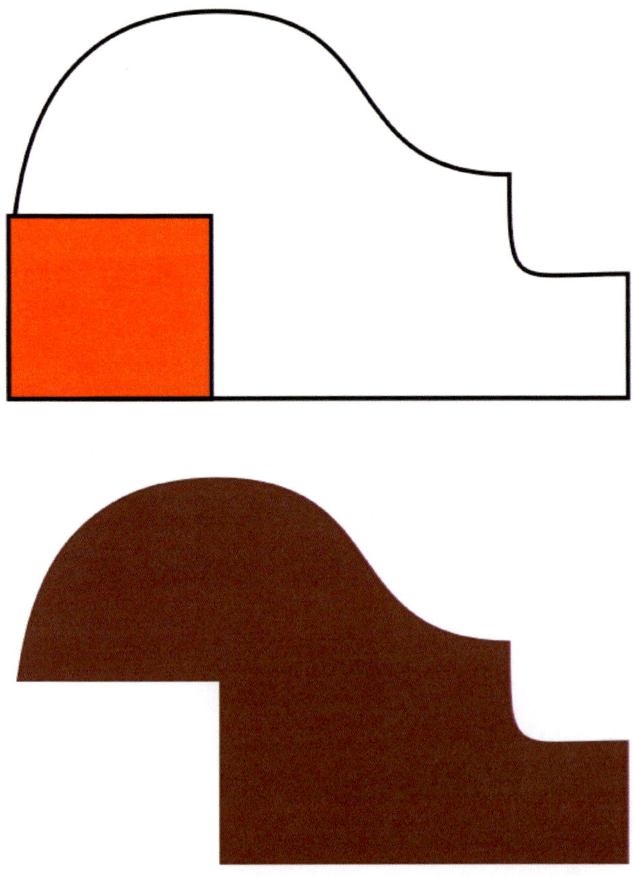

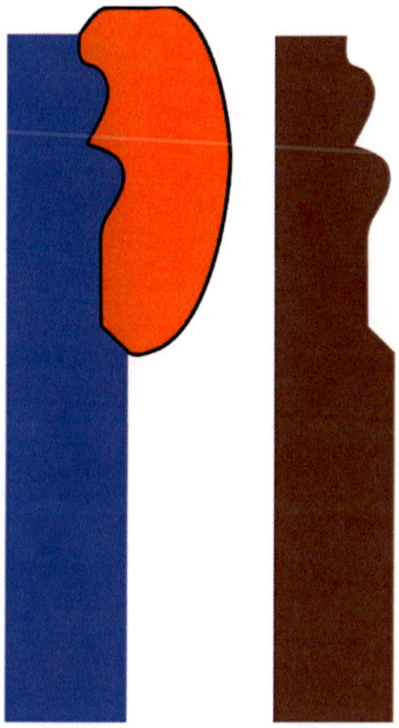

The skirting board I have done by the quick method - freehand drawing a shape (shown in red) to cut out of a rectangle using Path -> Difference in Inkscape. It may take you a couple of attempts to get it right and you can use the node tool to fix any minor imperfections.

The panelling is a simple rectangle with the corners cut off and the beading is a recess with a rectangle and circle in it. I have put 6 pieces together, with a dado at the top edge recessed into it and a skirting laid on it at the bottom.

I printed several of these sections to go on either side of the kitchen unit and stove.

## Coving

For the corner of a ceiling to the wall I made a piece of decorative trim.

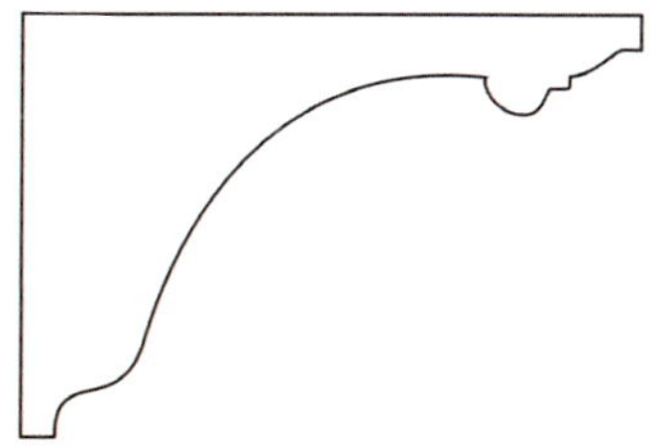

It can be as ornate or simple as you want. Victorian era items were rarely understated and so would have more detail, later periods were simpler.

I have also done an example of a simple equal sided curved shape.

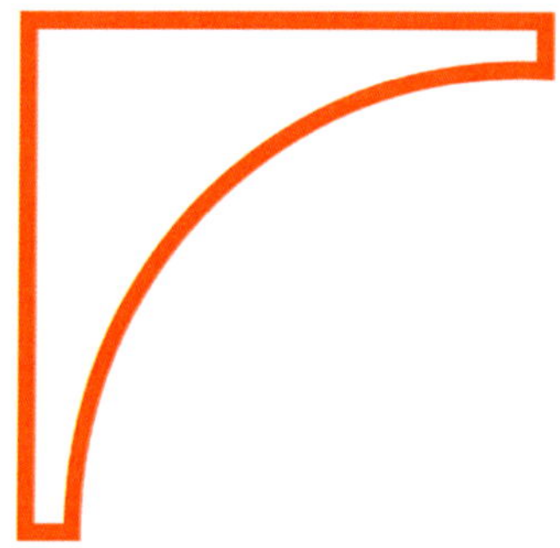

Print this on one side and as long as you need for the project, remember even if you have a small bed you can go diagonally to have less pieces and joins to cover.

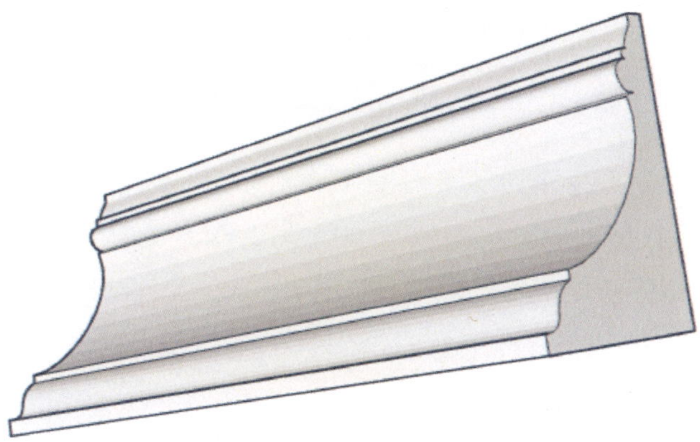

# Ceiling rose

How to design a ceiling rose.

Start with the size of your design, in full scale they are approx 1" (2 to 3 cm) deep but this can increase in the centre in some styles, 10" (254 mm) to 30" (762 mm) in diameter depending on the room size.

They can be just plain rings in more modern designs but the old ones will have detailed pattern work on them. I did a couple of example designs with a full detail outer circle and lesser detail inner circle, both re-using the design in a simpler form for the inner ring.

One design I took the Fleur De Lys from the tile, another I took a "scribble" from Tinkercad and added semi-circles.

Take a cylinder, add a slope up to the centre with a cut-off cone, and place a couple of torus rings at the outer and inner edge. These not only smooth the edge but also give you a guide to align patterns to.

Place your pattern and Duplicate and Repeat it with rotation in between until you have filled the rose. If it doesn't work the first time, undo and adjust until it does. It's best that it looks

perfect on screen as printing always adds it's own blurs and imperfections.

# Quick Guide to Tinkercad

## Viewpoints and zooming, moving

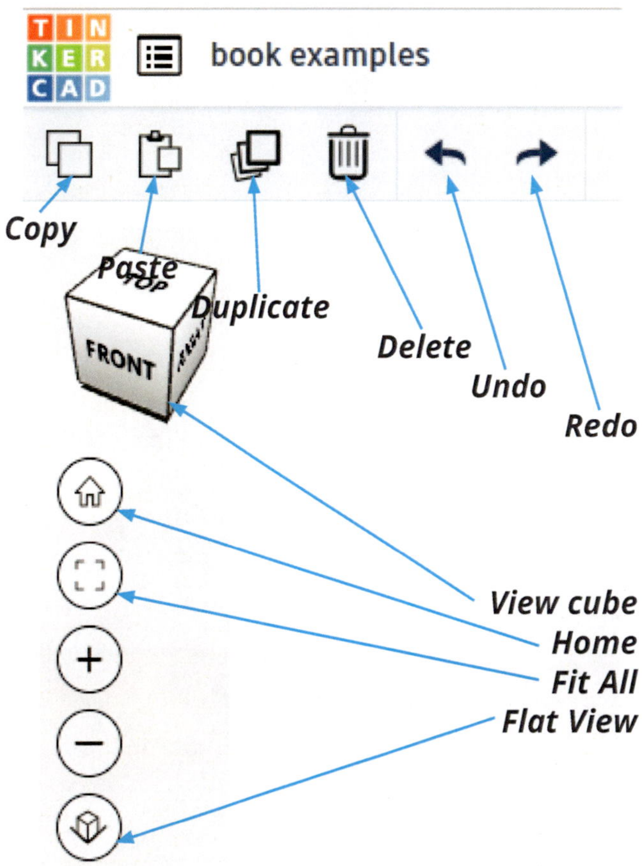


At the top next to the Tinkercad logo is the title of the current design (click to change) and the button to access your designs library.

Next is the icon panel with copy, paste, duplicate, delete, undo and redo.

Then is the view cube that you can move around with the mouse to adjust.

Under this is the home button to take you back to a view of the workspace.

Then the "corner box" (Fita ll in view) icon underneath will zoom you to a view of the selected object(s).

+ and - are zoom controls if you aren't using a mouse. Using the mouse wheel lets you zoom in and out.

Finally the Flat View icon will remove all the 3D representation which can get in th way of accurate positioning.

Holding the mouse wheel down lets you move the viewpoint across the workspace.

Practise these before doing an actual project to save yourself frustration.

## Primitives

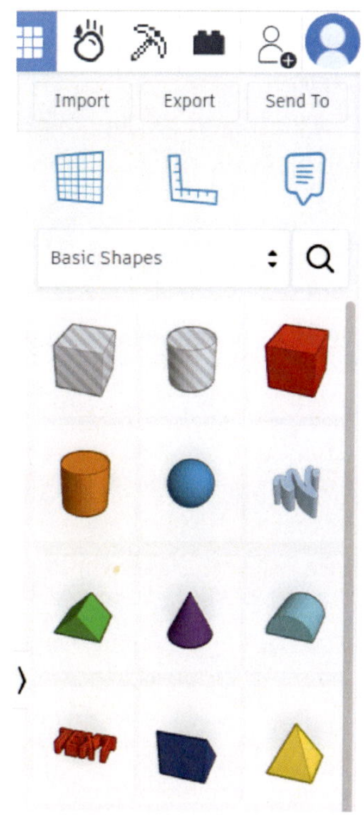


Primitives are simple cubes, cones, spheres etc. that you can use as building blocks in a design. In this book we will mostly stay on the Basic Shapes section as the others within it are generic clipart and specialist options.

Click and drag one of these shapes onto the workspace and then you can alter the parameters from its menu. An example is Cylinder which

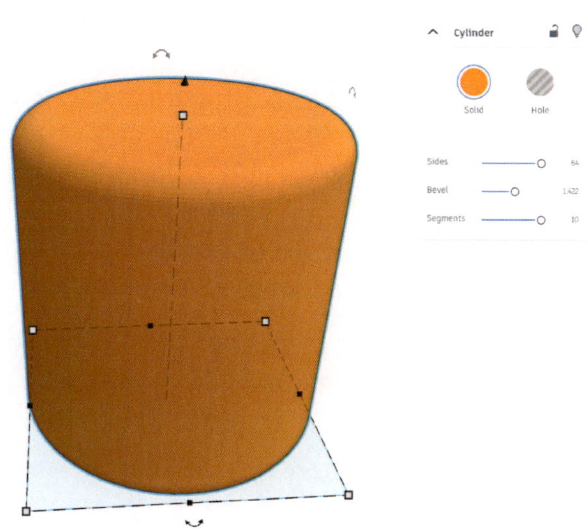

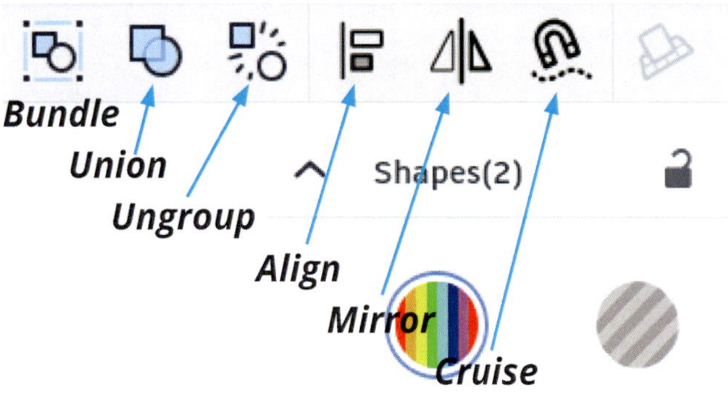


as well as adjusting diameter and height it has smoothing options. Bevel curves the hard edge down into the body of the cylinder. Sides smoothes the circular aspect of the cylinder, by default it is only 20 and can seem blocky if used large. Finally Segments smoothes the transition into the bevel.

## Grouping, aligning, flipping

You can combine objects together which is useful for moving connected objects around, also at this point holes will be cut from other objects to leave the final shape.

You can align objects to each other at the edges or centre of any plane.

You can mirror an object in any plane too, e.g. flip it back to front or upside down without rotation.

Finally is Cruise which is a kind of Snap-to mode for assembling objects.

## Import SVG

At the top right click Import, then find the SVG you want. It will be brought into the same place on the workboard as it is on your SVG design, allowing you to accurately overlay elements in multiple SVGs.

By default it is 10mm / 1cm high. If you click the white square in the centre you can then click the measurement 10.00 and type into it directly.

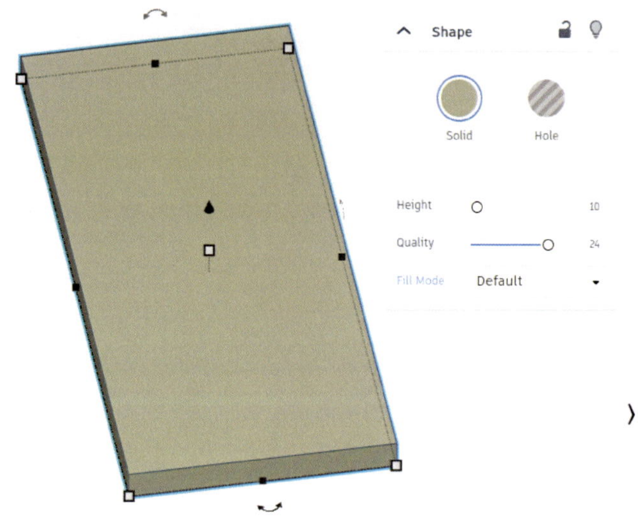


## Nudging

A simple concept - moving an object with the cursor keys. You can change how far it moves each time to suit, from 0.1mm to 5mm via the Snap Grid dropdown at the bottom right of the workspace.

## Height

The small black cone in the centre adjusts height from the baseline, use Snap Grid bottom right to control how finely you can move it (if you are lining up a feature by eye) or type directly into the box.

## Scaling

The white outer squares give you the dimensions of the sides and top, again if they need changing you can type directly into them.

## Rotate

There are 3 sets of double ended arrows when you click on an object, each offers rotation in a different plane. These can operate in 1 degree jumps at the outer ring or 22.5 degrees in the inner ring, where it is easier to get a 45 or 90 degree turn.

## Duplicate with operation

Using the Duplicate & Repeat feature you can make a copy, it will be exactly in the place of the first one. Then you can do a move, scale, rotate etc. and duplicate again and it will repeat the operations on each duplicate with precision. We use this in the vase project.

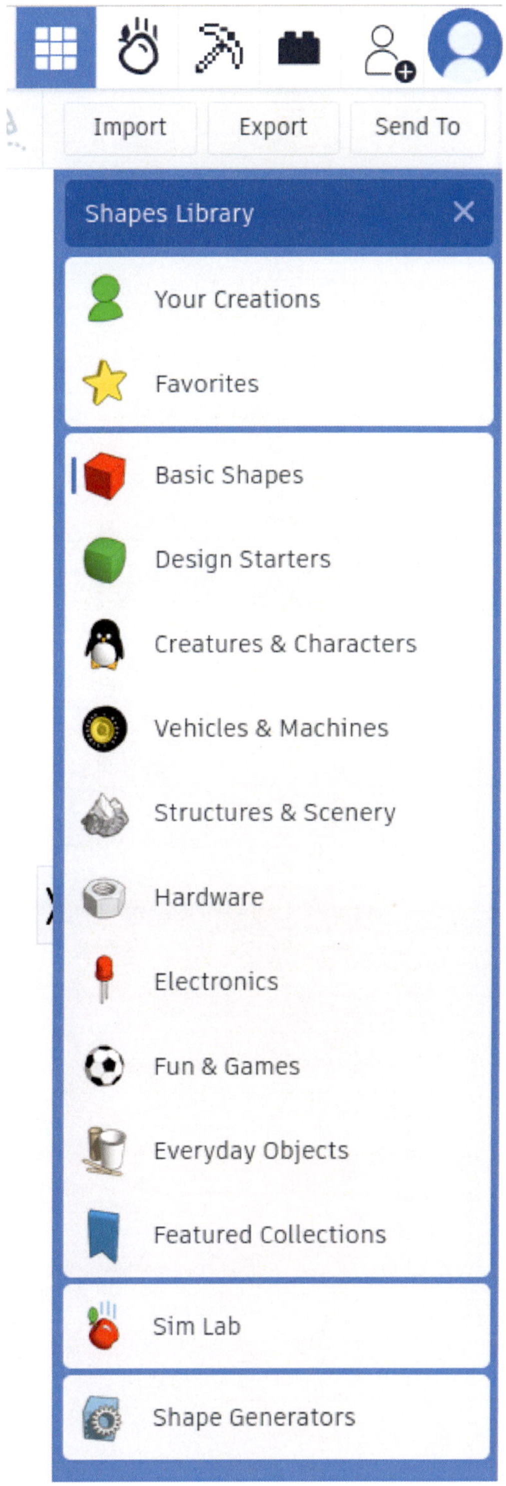


## Create shape

Sometimes there isn't a primitive for what you want and you can model it yourself. You can cobble together whatever you need from primitives and SVGs until it is right, then go to Your Creations -> Create Shape. The shape will be welded and simplified and stored as a new primitive. We used it in the vase project (book #1) to turn 180 shapes into one, simple, whole.

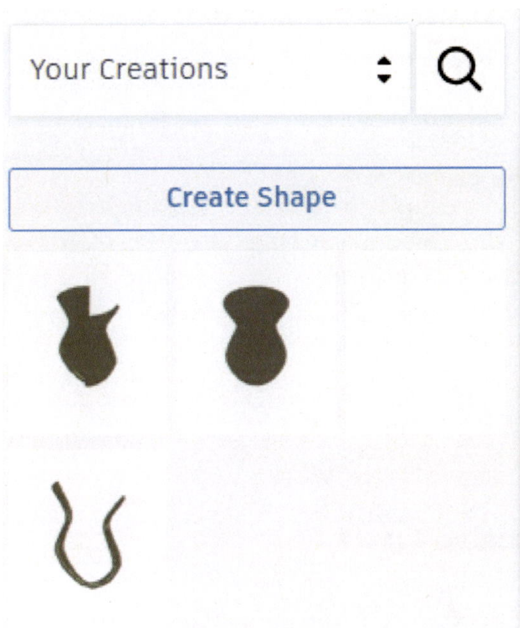


## Scaling

If you want to make a version for 24th scale the easiest way is to select it all, hold shift, click on the corner white square and type 50% in one of the measurements. All sides will then scale at once.

*The official tutorials are at* ***www.tinkercad.com/learn***

# Quick Guide to Inkscape

## Curves, paths, nodes

For the vase project in book #1 we needed to draw a curved shape that is perfectly symmetrical. Straight lines are easy but how to turn them into curves?

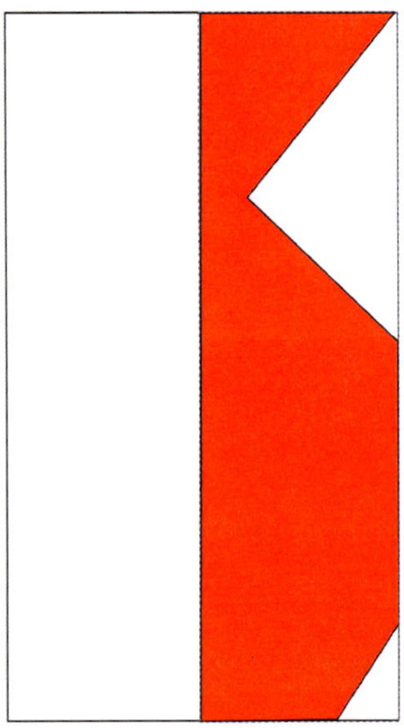

First I draw a box the total width and height I want, then another box exactly 50% of the width.

With the line tool draw the vague shape you want.

To add curves click on your object, use the Node Tool, click where you want the curve and on the top bar click "Add curve

handles". This will now give you two circles either side of the point that you can move freely until it turns into the shape you want.

## Nudging

Moving an object with the cursor keys - you can choose how much from Preferences -> Behaviour -> Steps

## Snapping

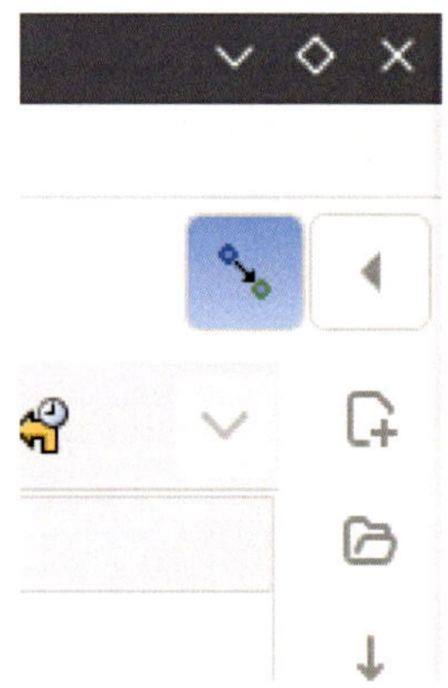

This is a useful assistant in making things fit perfectly. You can move an object to be exactly at the join of another. You have to toggle this on or off via the top control bar, hover over the icons until you find Snapping.

## Group

A group is a collection of objects kept together, they move together and scale together. This helps keep relative positions intact, click Object -> Group.

## Weld / union, trim, intersect etc.

Welding, or Union in Inkscape, is joining objects into one. Different to grouping, in that if you take a group to a cutting machine it will cut around all the outlines of the object, with Union there will only be the outline of the whole.

Select all the items, check they are ungrouped, and click Path -> Union. (There is an example later under Duplicate.)

To cut a shape out of another shape use Path -> Difference.

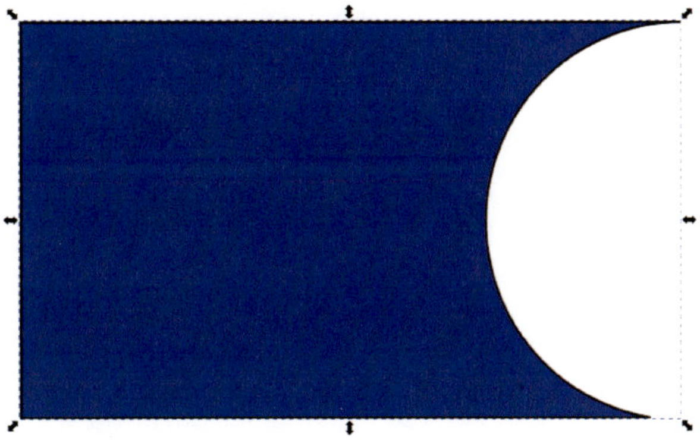

To get only the parts that cross each other use Path -> Intersection.

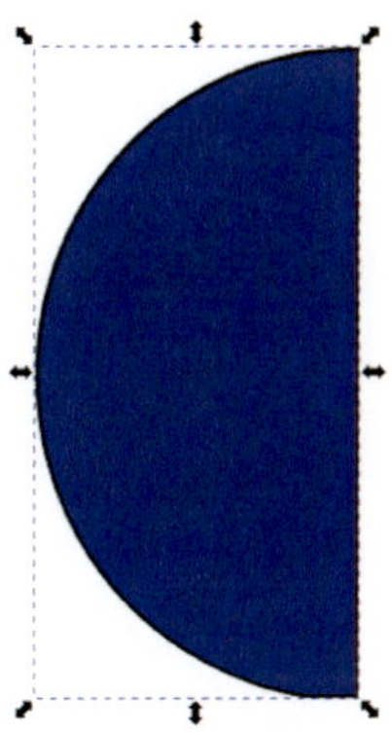

Some of this might not seem useful at the moment but quickly practise them as they are fundamental building blocks in design.

## Closing paths

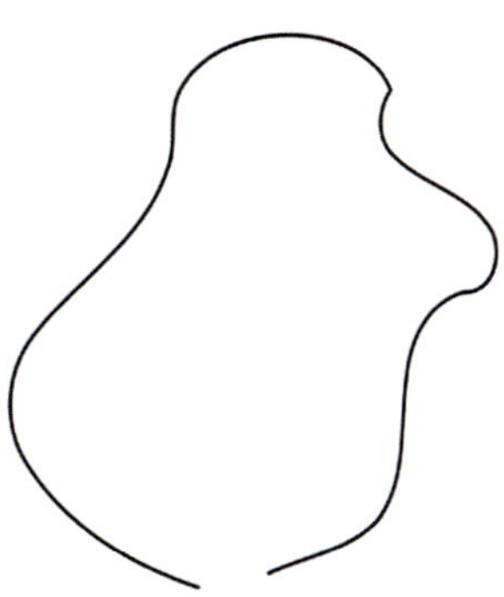

If you draw a shape and it doesn't quite meet up you won't be able to fill it or use it in Tinkercad. Select it and click Path -> Union.

It is auto-completed and can be filled and used.

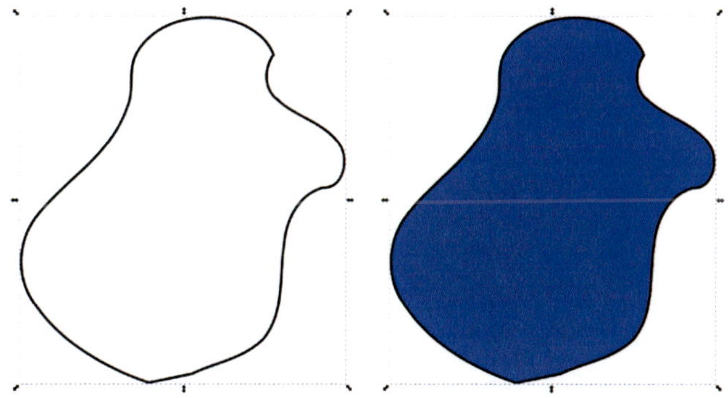

## Tracing

Sometimes designs are so complex you will find it easier to sketch it freehand than make it onscreen using the curve tools. That's OK as most software has a trace option that will do its best to draw the outlines for you. You can then fine tune them until it is exactly what you want.

File -> Import and choose your image.

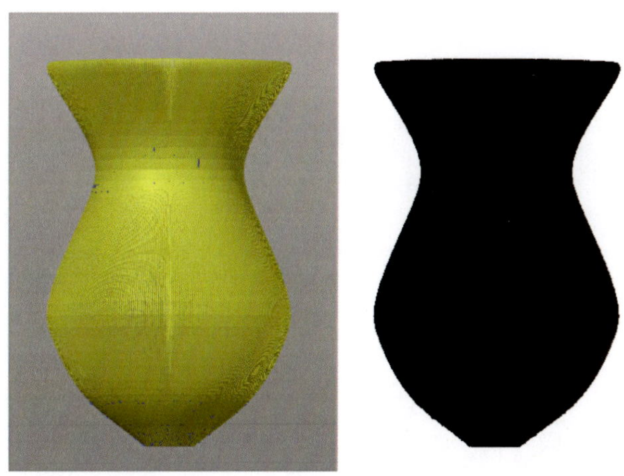

For this example due to copyright reasons I have chosen one of the self-generated pictures from book #1, then tweaked it with a filter to get the best contrast.

At the top right panel I chose Trace Bitmap (another name for an image), then went through the options to see what it could do for me nearest to what I wanted.

Edge Detection gave me an object made of a double line where I chose the thickness between the two, which is an ideal start. Move the original image away. All I have to do now

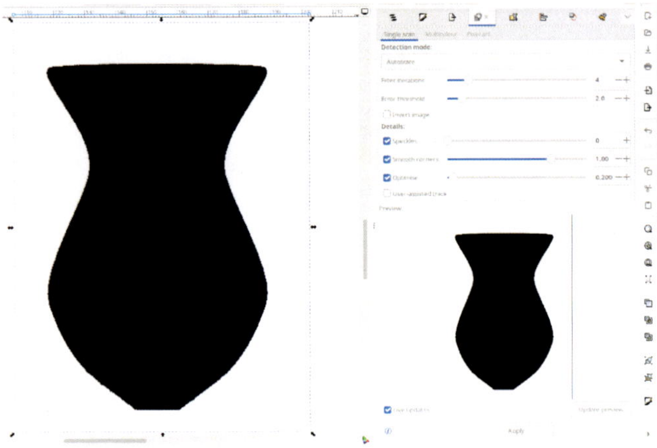

is edit the top using the node tool and I have an outline I can use.

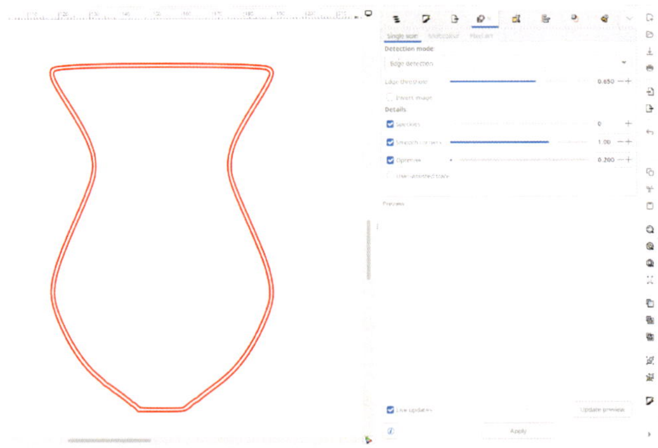

## Duplicate (showing how to make a cog)

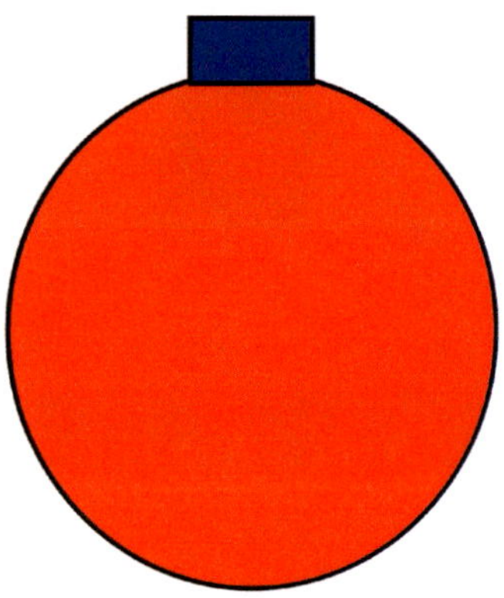

More than just copy and paste, duplicating with a repeated operation lets you create new shapes. As an example I shall make a basic cog by making a circle with a box on top. Also add a tiny circle centred on the first circle. Click on the box twice and a cross should appear - this is the centre of rotation. Drag it to the tiny circle at the centre. Now as you copy and rotate the box it will become a pivot to move the new boxes around the circle.

At the right hand side of the screen click the Transform icon, then rotate and put 45 degrees.

Now Duplicate from the Edit menu and Apply on the rotation box. When you next Duplicate it will keep rotating by 45 Degrees until the circle is full of cog teeth.

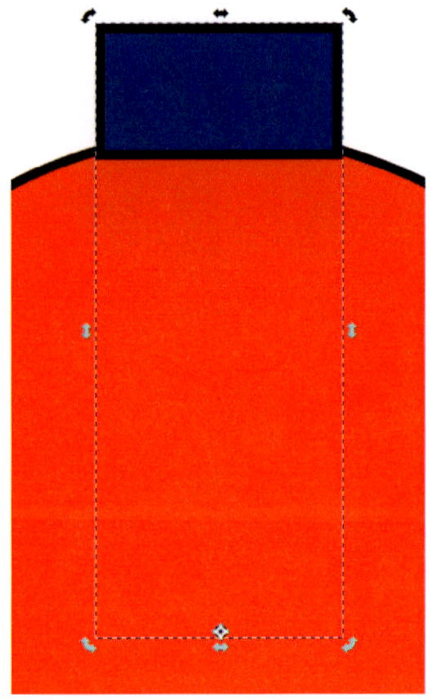

It is important to check that things went exactly to plan and touch where you want before the next

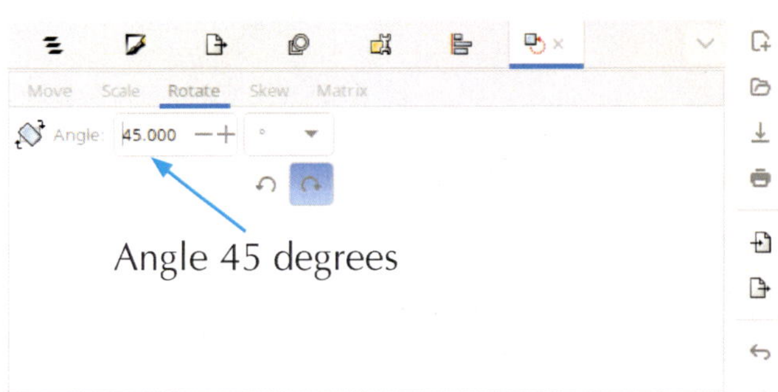


step so go to View -> Display Mode -> Outline and you will see that the boxes overlap into the circle, adjust them now if not.

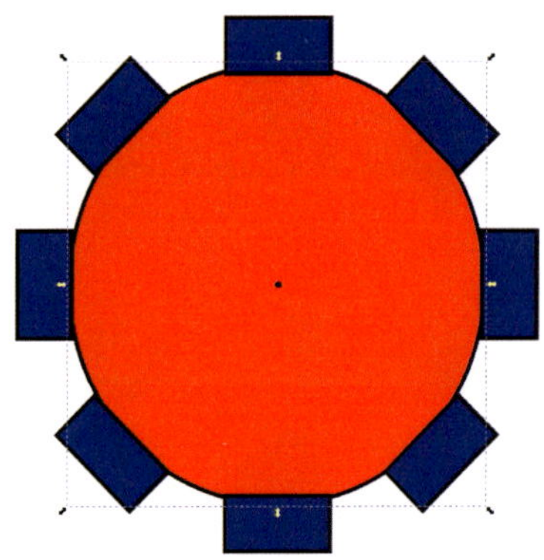

Select the circle and teeth and click Path -> Union. You should now see an outline of a cog without the component boxes.

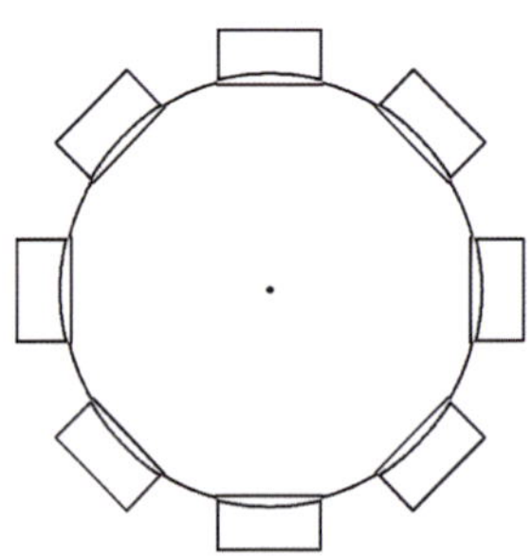

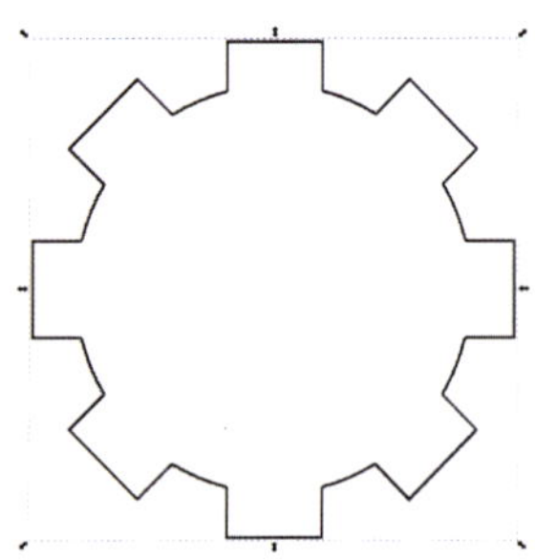

# Align

This is available from the Align and Distribute button at the top right.  Select two or more objects then chose to have them aligned vertically or horizontally using the buttons.

# Distribute

From the same place you can choose to have items equally spaced either horizontally or vertically.

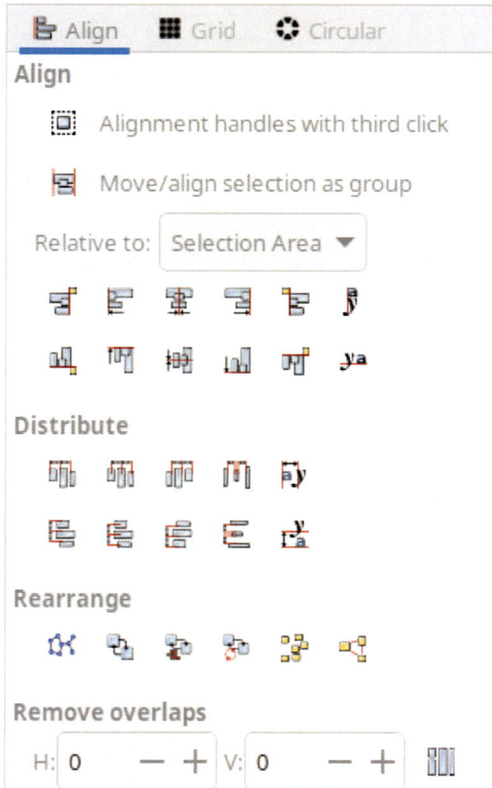


*Inkscape is a powerful package available on many desktop platforms and has extensive tutorials available online, here I have just covered the operations most used in this book.*

*The official tutorials are at* ***inkscape.org/learn/tutorials/***

# Quick Guide to Prusaslicer

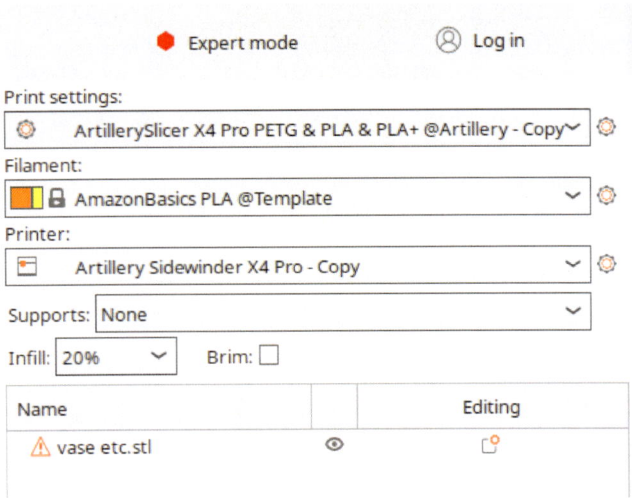


## Print settings

You might ask why you need multiple settings as you only have one printer, but most printers come with lots of presets already set up here e.g. "0.5mm layer height, ultra detail, slow", "0.3mm layer height draft, speed" etc. You can create and save these off for your own printer if they are not supplied

## Select material

At the top right you can choose which material you are going to print with. There is a library in the slicer with common filaments that knows their printing and bed temperatures and characteristics.

## Select printer

Also there is a dropdown box for printers. The configuration wizard lets you choose the default entry here for new projects.

## Add STL to cut mat

Unless you have come straight into the slicer from Tinkercad, the first thing you need to do is put a model on the cut mat. You can do this with the Add button or File -> Import-> STL.

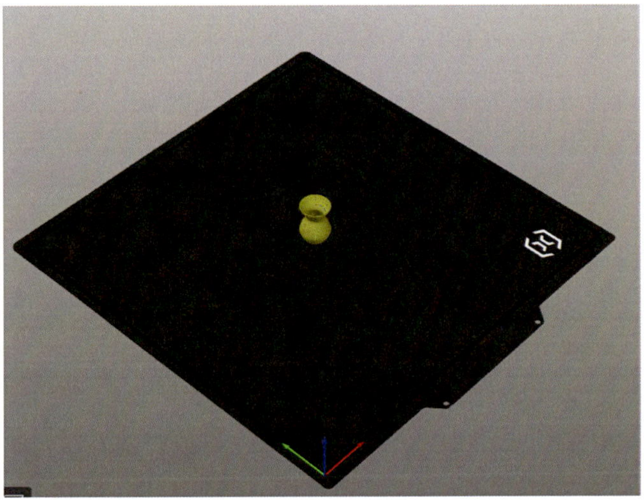

## Arrange

When you add more than one item to the mat it might end up with them clashing or even with them all in the same place at the centre. The Arrange button will space them out evenly and sensibly across the space.

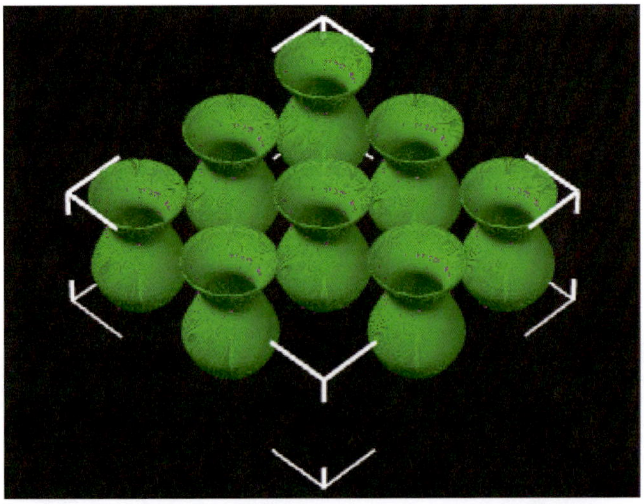

# Scale

When I am first working on a project I print all my tests in 24th scale at draft speed. Click on the object, click the Scale button at the left and either manually drag it to approximate size or type 50 into the Scale Factors box at the bottom left.

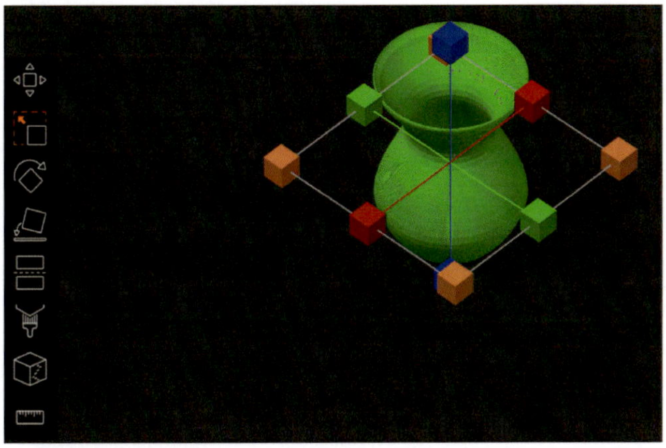

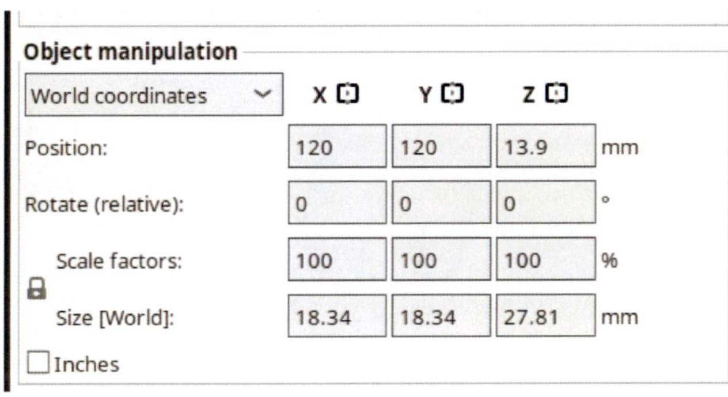


# Slice

At the bottom right click Slice Now, it will render up the model to G-code. Then it will provide a panel with estimates of time needed and material usage to give you some idea. These estimates are not perfect but they are a good ballpark figure. The estimates on the printer during printing are often less accurate.

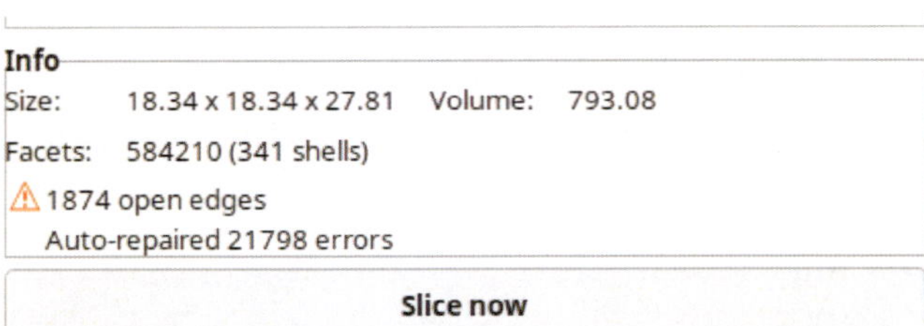


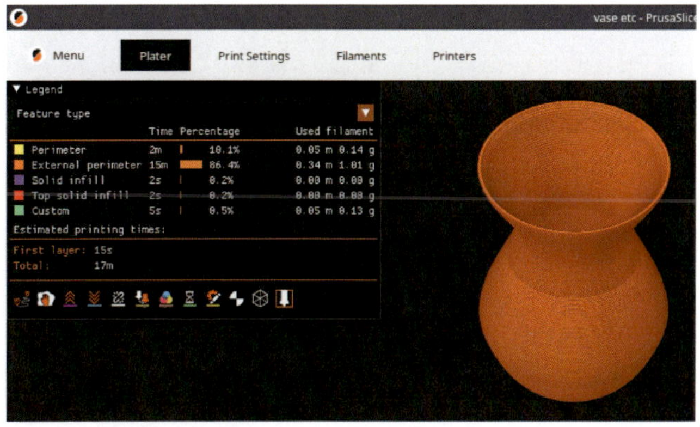


# Export G-code

At the bottom right click Export G-code to save off the file. It will be named after the first STL, the type of material, and the estimated time, e.g. vase_PLA_0h17m.gcode

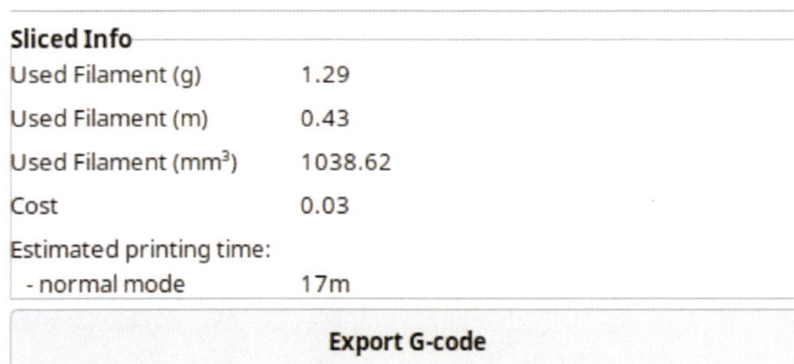


# Take to printer

When you have saved the g-code file, either copy it to USB or SD card, or copy it to the printer via Wifi, whichever method your printer uses.

# Viewer

If you have badly named some files and are not sure what they will produce, Prusa GCode viewer will show you the contents. This is handy on basic printers with no display panel.

*The official tutorials are at*
***help.prusa3d.com***
*and click PrusaSlicer documentation*

# Quick Guide to Bambu Studio

*Bambu Studio is based on Prusaslicer, but with added capabilities for the Bambu printers and network capabilities.*

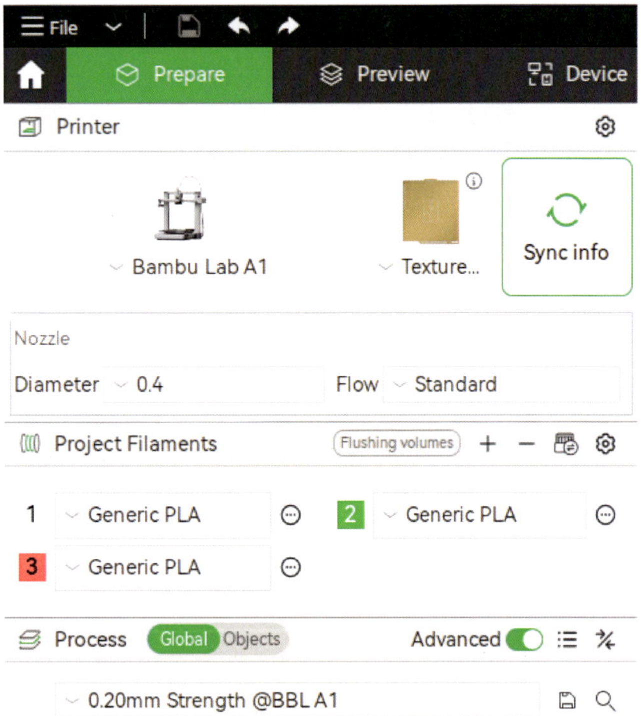


First you can choose the filaments you want to use in the project from the multicolour unit.

If your printer is turned on it will sync the choices from it, as long as you remember to edit the colour if you add in a roll of generic filament. Bambu filaments should be detected automatically.

Next you can assign each object in the print a different colour - I choose green for the leaves, terracotta for the vase and petals, white for the tiles.

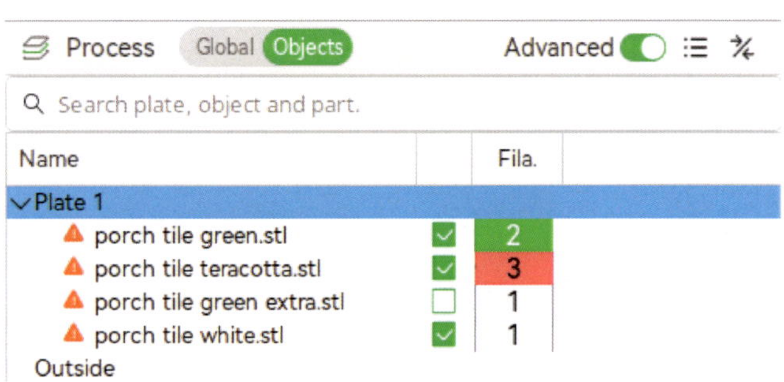


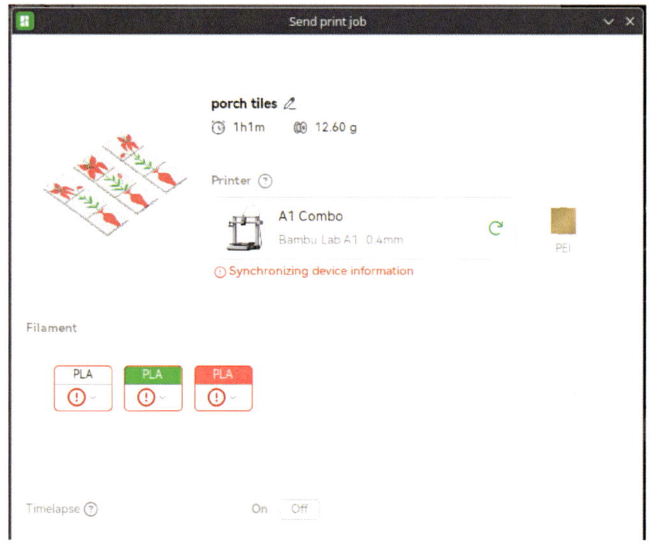


Click preview at the top and it will give you a summary of times and filament usage. If you have no warnings and are happy click Print Plate to send it through the cloud.

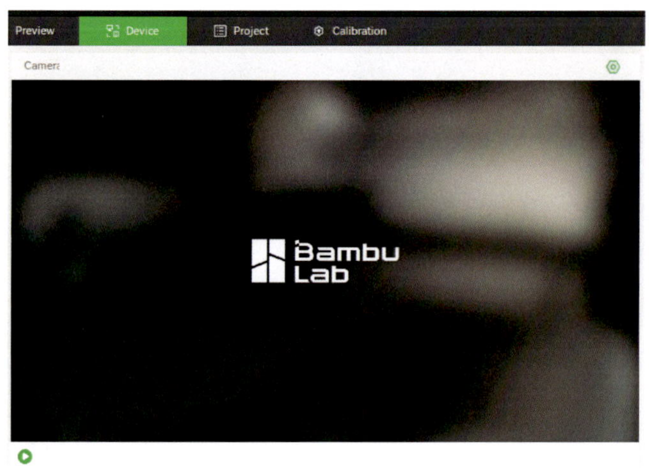


On your computer or on the app you can preview the print with this video window, just click the Play button.

By default the lighting is poor and badly positioned for miniatures on a Bambu A1 but I added lighting quite cheaply. I may add a Raspberry Pi with a webcam and AI print monitoring later.

# Appendix - 3D Printing Basics & Slicing

## Printer parts

The machine is normally one of two types - filament or resin. Here is a quick guide to the important pieces.

### A filament printer will have:

**Controller / circuit board** - where the movement instructions are processed (internal)

**X/Y/Z motors** - to move the bed (normally Y - front to back) and extruder (normally X - side to side,Z - up and down)

**USB / SD / Wifi** - to transfer printer files

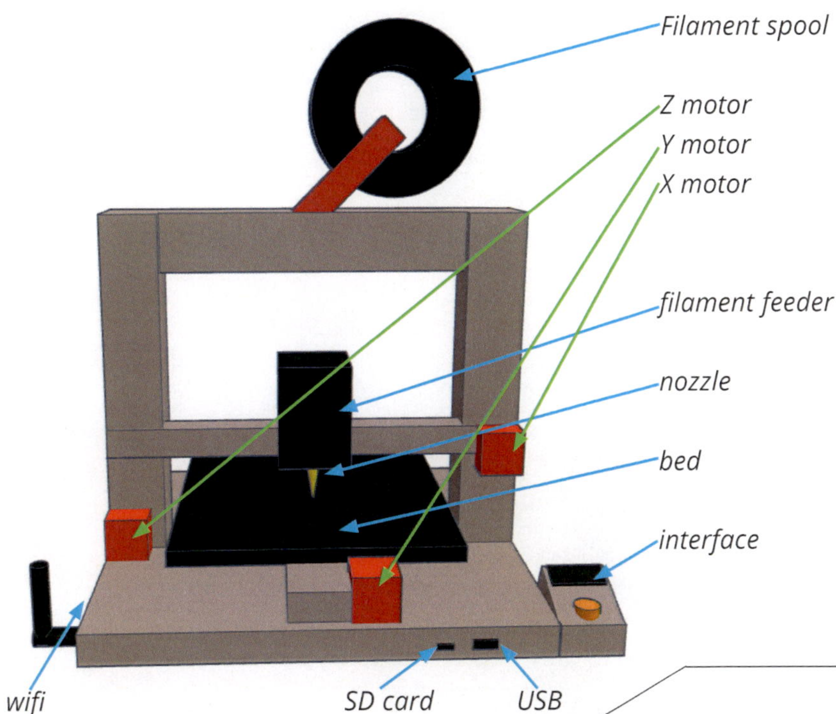


**Interface** - a dial and buttons or a touchscreen to control the printer

**Filament** - the material (e.g. PLA) used to make the prints

There will also be **fans** to cool electronics and filament, and **structures** to hold all these parts together.

**Extruder** - the hot part of the machine that moves and ejects the filament

**Nozzle** - the part the filament is squeezed out of, it can come in varying widths.

**Bed** - the base for printing onto, normally heated

**Filament feeder** - where the spool of filament is pulled into the extruder

### A resin printer will have:

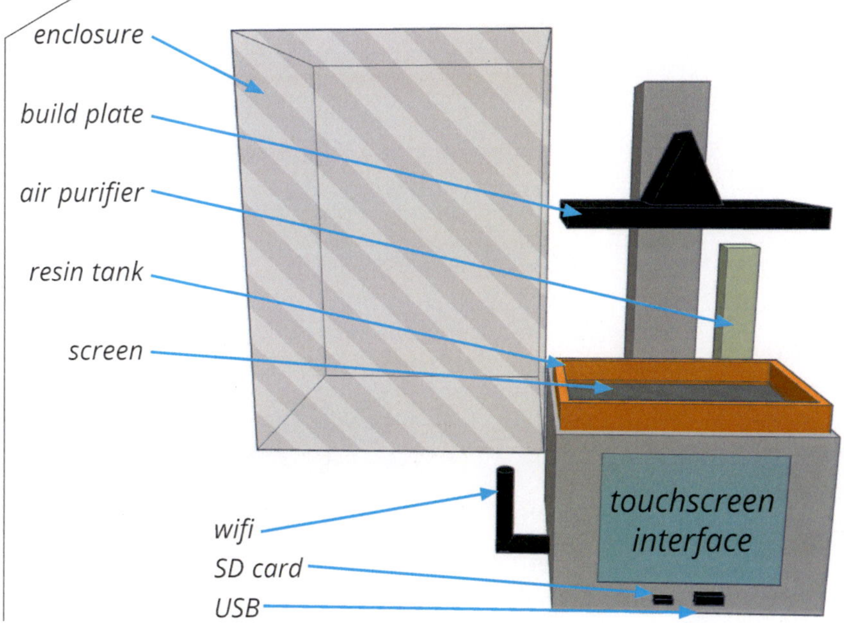

**Screen** - to project a slice of the image onto the liquid

**Bed or build plate** - to hold the print, which is lifted vertically away from the screen at each stage

**Tank** - to hold resin near the screen

Film - transparent plastic between the resin tank and the screen

**Controller / circuit board** - where the movement instructions are processed (internal)

**Z motor** - to move the build plate up and down (possibly internal)

**USB / SD / Wifi** - to transfer printer files

**Interface** - a dial and buttons or a touchscreen to control the printer

**Resin** - the light sensitive liquid

**Curer (optional extra)** - to set the model firmly after print

**Washer (optional extra)** - to clean off excessive resin

A resin printer may also have **air filters** as the fumes from them can be toxic depending on the resin used, and will be in an **enclosure** that blocks light.

## Modelling

You need to download or design something to print (see design section). There are free models and free design software out there already.

## Slicing

Starting with a free slicer like Cura or PrusaSlicer worked well for me and I haven't felt the need to upgrade to anything else yet.

There is always setup time for these programs, where you tell it the machine you have and the limits it has to print. For instance I have a Geeetech A10 with a 22cmx22cm quoted print bed area and a 26cm maximum height. The height is apparently optimistic and prone to crunching the cables so I never go past 22cm.

Next you select the filament you will use (PLA, PLA+, PETG etc.), normally there is a handy list of presets to make this just a tick from a selection, but if something is a little different it is possible to tweak this too.

Finally, the fine tuning of settings. I have made a lot of changes after finding various small issues like stringing issues (lots of messy strings on your print) and bridging issues (printing over holes).

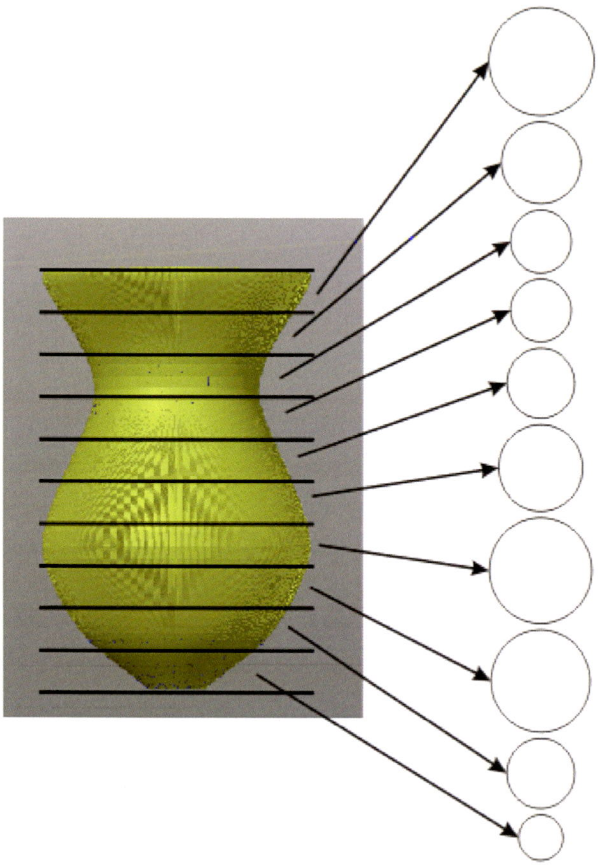

What is slicing? The software takes your 3D model and cuts it into sections to print it, in the simplified vase example it is turned into a succession of rings. Starting at the bottom the printer will take each ring and print it then move up a step at a time. In reality the printer will have many thousands of these steps at 0.1mm, which is the finest for filament, potentially even finer for resin.

It relies on the heat to glue each band together, then quickly cool to set as a solid object. However this can leave a line as it stops at each section, instead you can activate Spiral Vase mode to keep it moving around and up at all times.

## Bed levelling

You must do this first and check it regularly on some printers. It is normally simple and sometimes automated. My printer is half automated - it moves to the corners but I adjust a knob manually so the head just scrapes a sheet of paper pulled over the bed. After setting all 4 corners do it again as there is often a change that affects the first corner.

You can get free test prints that will tell you if you are good to go such as
**www.thingiverse.com/thing:1557995**
*"Bed Level Calibration by zumili"*

## Test print

When you unpack and set up the machine make sure to print the supplied test print. Of 3 machines, the 2 filament ones never made it to the end and I had to tune them in some way. The resin one printed it perfectly but it was absolutely welded to the bed so I knew that future prints needed a "raft" if they were delicate.

## Resin printing extra considerations

Resin printers submerge a platform in liquid as it prints, there is always a pool to be accounted for as you form a shape. Some designs can be hollowed out to save resin and this can be done automatically. You will probably need to add drainage holes too or raw resin will be trapped inside the print, again this would be costly. Remember the print is upside down and to drain towards the top. Experienced users recommend printing on a raft with supports, sometimes with the object at the best angle to accommodate them.

## Filament infill

With filament it is normally pretty simple, you can just reduce the amount of Infill to suit, 10-20% on a large object often still leaves enough strength to be fine. Smaller pieces that may have some strain on them can go between 50% up to 100% for very small pieces.

I did have issues with large flat surfaces and 10% infill, though this might have been an issue with my printer which failed shortly afterwards.

# Appendix - sizes

*A guide to dimensions for average full size architectural items. Note they can vary by period.*

## Stools

Counter stools 59 - 70 cm (23 - 27″) tall, bar stools 71 - 84 cm (28 - 33″) tall

For a 92 cm (36″) counter you want a 61 cm (24″) bar stool.

For a 107 cm (42″) bar you need a 74 - 77 cm (29 - 30″) inch stool

## Chairs

They should be 44 - 60 cm (17 - 23″) high to the seat for chairs or benches. They should leave leg room under any table or counter they are used at.

## Tables

Dining table height is normally 71 - 76 cm (28 - 30″), width is 92 - 107 cm (36 - 42″), the length to seat four should be 122 cm (48″) long. Tables for four to six people should be 132 cm (60″) long. Tables that seat six to eight people should be 198 cm (78″) long.

## Counters (e.g. bar)

The bar height is normally 107 cm (42″) from floor to countertop.

## Kitchen units

Modern height for a kitchen counter is typically 91.4 cm (36″) from the floor. Historic height is 76-84 cm (30-33″). Pantry cabinets can be up to 244 cm (8′).

Normal depth is 60cm - 65cm (23-26″).

## Bay windows

Widths typically range between 150 cm and 300 cm (59 - 118″) - measure across the widest part of the opening outer frame.

Height of a bay window is generally 120 cm to 150 cm (47 - 59″) - measure from the sill to the top of the frame

135° and 150° angles are normal in Victorian or Edwardian houses.

Sill heights are often around 90 cm (3′). Internal window seat height of 45 cm (18″), 40 - 52 cm (16 - 20″) in depth.

## Door

Standard front door sizes can range from 76 - 115 cm (30 - 45″) in width and 180 - 244 cm (71 - 96″) in height. Typically they are 92 cm (36″) wide by 204 cm (80″) tall.

Internal doors are 198 - 205 cm (78 - 80″) high. Typical width is 76 cm (30″).

## Columns

Victorian styles use porch columns that are 244 cm (8′) high, 10 - 15 cm (4 - 6″) in diameter if milled.

Round columns have diameters 15 - 36 cm (6 - 14″), proportionate to the porch height.

The diameter of a column should be about 1/9th of its height.

## Coving

Size depends on ceiling height and room size, ranging from 5 - 7 cm (2 - 3″) drop to 12 cm+ (5″+) drop. For a standard 2.4m (8′) ceiling, 7 cm (3″) drop is recommended.

## Dado rail

Positioned at a height of approx 1m (39″). Originally to protect a wall against chair backs.

## Ceiling rose

They are 2 - 3 cm (1″) deep, 25 cm (10″) to 76 cm (30″) indiameter dependent on room size. Use 45 cm (18″) as average.

## Bricks by era

Victorian bricks were approx 228 mm (92) wide by 114 mm (4.5″) and 74 mm (3″) high. The height can be a little more and up to 25 mm (1″) less depending on the era. It is worth double checking before making a period piece.

# Other Software

*A quick mention of other common software recommended by 3D printer users.*

## Blender (Win/Mac/Linux)

Free, very comprehensive animation and 3D modelling software, plenty of free tutorials available. Note it has a very steep learning curve.

**https://www.blender.org**

## Freecad (Win/Mac/Linux)

Quite comprehensive CAD software, plenty of free tutorials available.

**https://www.freecad.org**

## Fusion (Web or Win/Mac)

From Autodesk the people who offer Tinkercad but different, with many more options. It has a free license for non-commercial use.

**https://www.autodesk.com/products/fusion-360/overview**

## Nomad Sculpt (Win/Mac/Android/iOS)

Especially good for use on tablets with a stylus. Relatively cheap licensing.

**https://nomadsculpt.com**

## Onshape (web)

It has a free license for non-commercial use. Comprehensive CAD software.

**https://www.onshape.com**

*And lastly a piece of utility software that is often used for enlarging small models to huge sizes.*

## Luban (Win/Mac/Linux)

This is often what people use for full size character models.

**https://www.luban3d.com**

# Errors & questions

## How to ask for support online

If you want a chance of getting an answer to your question first search the group or forum to see if they have answered it already.

Next if it is a general 3D printing group they need to know your printer, what filament you are using, possibly the filament and bed temperatures you used as a minimum.

If you include a photo make sure it is in focus and well lit.

What not to do: an example I see five times a day in one small group on Facebook alone **"[dark blurry photo] - WHY IS IT DOING THIS?"** Why is what doing what? is the normal reply by exasperated people there to help if they can.

## Temperature too low

After changing filament temperatures in the slicer I accidentally typed the PLA first layer temperature as 20 not 220 and immediately started having issues. The Artillery Sidewinder just stopped after warmup, no clue why via the monitoring software. I assumed it had another issue and swapped to my reserve printer. The Geeetech has a much simpler two line text display and it says WHY!

## Damp filament

Depending where you live and the time of year you might start getting issues with the filament being damp, causing misprints, blobs and generally not working well. If you are in a humid place or using certain filaments that need to be dried before use, you can get a filament dryer. For some machines you can feed directly from the dryer to the printer.

## Holes or solids in Tinkercad

When slicing up solids to make shapes that you later use as holes Tinkercad sometimes gets confused. The solution is to not group holes together and not group items made with holes until the very last stage. I tried using "Create Shape" to simplify the structure into one mesh but had the same problems.

## Not at zero

If things make a stringy mess and nothing fixes it, it might just be that one of the objects isn't at zero on the Z axis. This makes the printer print in mid air, and the filament cools before touching the bed so that it doesn't stick.

## Head crash

A pile of spaghetti followed by a head crash on restarting. The machine just pushes into the bed to the full extent of the springs and gives an error instead of levelling.

It's likely the Z sensor in most printers.

However in my Artillery X4 in this case it was a design fault - the hot end is held in place by a tiny screw inside the case. This screw is pushed against 200 times every print as the machine probes the level of the bed. Eventually it works loose and the Z sensor can't get a reliable reading for the bed height. Apparently the only cure is regular checking and use thread locker paste.

## "Juddering" X axis

You can check the tightness of the belt. Maybe it needs some lube because it is catching as it goes around. Maybe there is a motor error, it can be possible to swap the motors temporarily to find out. Finally the motherboard or motor may have a loose connector.

## My print stopped halfway!

If it took a long time to get this far you might be able to rescue it by measuring the height, cut the model in the slicer and print the top bit to glue it on.

## Big blob print attached to printer, help!

You need to use heat to remove - a hair dryer or heat gun carefully pointed away from delicate wires. The cause is often the model slipping off the bed during printing, you can prevent this by keeping the bed spotless using washing up liquid or by applying glue to it. I use both on different machines.

If you touch the bed constantly getting prints off it will bite you later if you don't clean or glue.

## Prime tower - why - how/should I turn it off?

This column makes sure the changeover of filaments is perfect. This can be a time waster and filament waster but do you need it? You probably do for delicate patterns on miniatures. For structural pieces you can turn it off.

## Can I use ipad/phone instead of desktop?

Yes, e.g. for Tinkercad you can design and Bambu Handy you can start and check live video of prints. If you get more into design a cheap computer will help enormously, it doesn't need a lot of power as Tinkercad runs on the web and the slicer is only run to generate a print so a little patience can save a lot of money. The type of computer or laptop (Windows / Mac / Linux) isn't normally important except Chromebooks may have issues that you need to check before buying.

## How much to run a printer?

The electricity element of printing is not so bad, after warming the bed and hotend the printer may run at only 150W or so continuous power draw.

Filament is **$€£** 18 or so per kg and for miniatures it will last a long time.

## Is it food safe?

It can be, if you check the filament and type carefully, but it is not necessarily waterproof so may need to seal too. Again you need to check the sealant for food safe too.

## Should I vent the room?

Some people sleep next to printers running overnight. I don't advise it for nise alone, never mind potential fumes. PLA, ABS, resin etc. all give off something - ABS and resin particularly so - so if possible air the space and don't breathe in for long periods.

## Should I use glue on the bed?

Or wash with soap and water? Or use Hairspray? It's personal choice, I ran one machine washed and one with glue for a few months. It is easy to see when the glue has worn but almost impossible to see if you touched and left a smudge of grease on a bed, so the washing approach works well with discipline but can give nasty surprises.

## Will AI make a model for me?

Maybe ... but often not. It can be OK for making cartoon-ish figures that other software then makes a model from.

## Can I print to sell? Can I make money from it?

It's complicated ... there is now lots of competition online and at fairs, licensing can cost you a monthly subscription and not guarantee other people won't be selling the same thing next to you. It is far easier with your own, new, designs.

## Make supports easy to remove?

If you have multicolour use PLA with PETG, whichever is the main body use the other material for the contact point of the supports. This makes them easier and cleaner to remove.

## Which printer should I buy?

By budget, ease of use, type of miniature (resin for detail, filament for larger items) You can always start with a Bambu for filament and Elegoo for resin. There are fans and detractors for any machine but the Bambu cetainly is far simpler for a beginner than anything else near the price.

## Do you leave your printers unattended?

If you do it is essential to have remote monitoring by video and I recommend a remote power switch so you can pull the power from the system if it is looking bad.

The first book in the series, focussing on an old shop and its fittings, display units, cash register and stands.

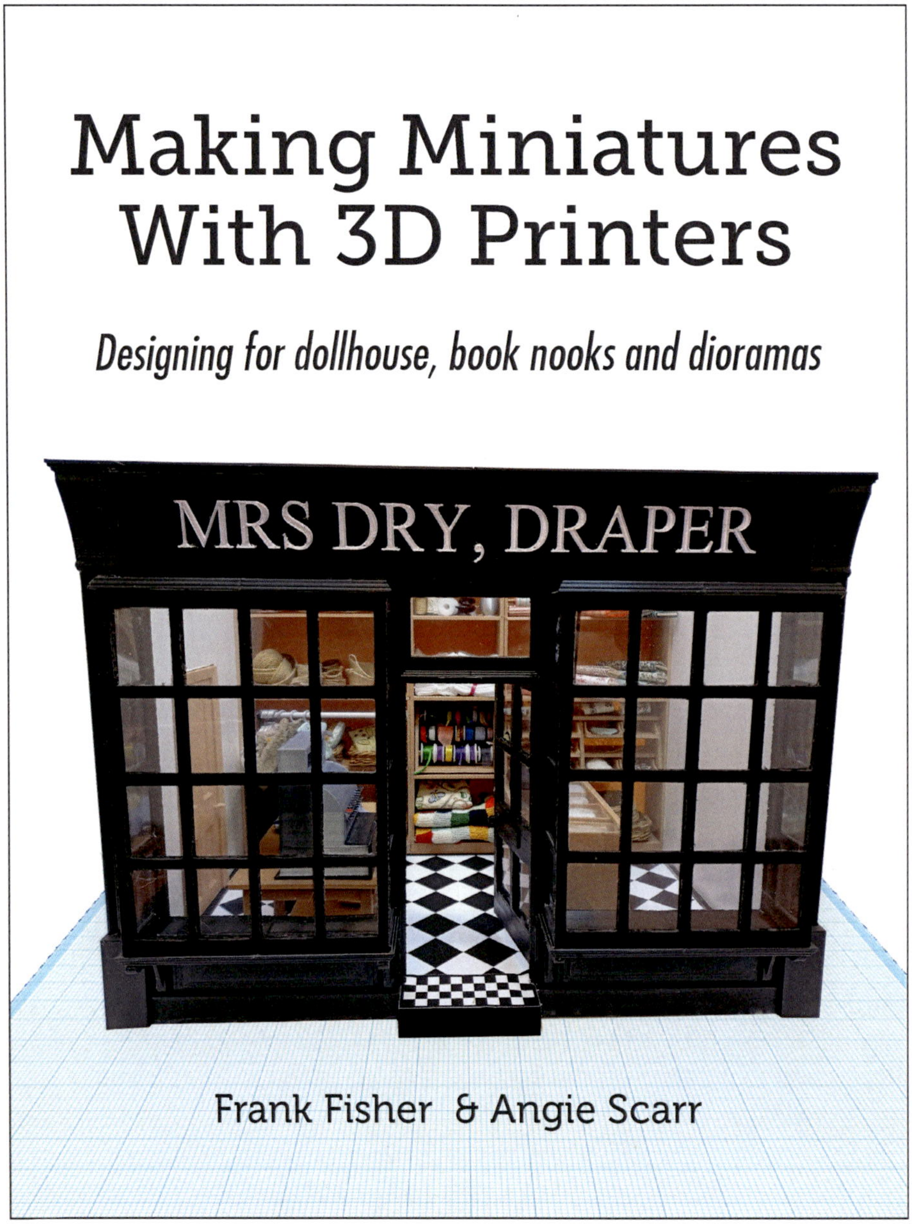


## Available from Amazon on ISBN print 979-8301515491

released on Amazon 2024

Indie bookshops ask for ISBN 978-841260238

ebook https://books2read.com/u/mexn5R

## Biography

Frank Fisher is best described as a 'creative technician'. Since the very early days of the internet he has always helped musicians, artists and writers do whatever it took to publish or reproduce their work. Producing websites, music, videos and books, along the way doing all the IT work it took to turn those creatives dreams into physical and saleable forms. Frank fills in the holes in the IT knowledge of his clients and often his friends and family. His love of and ability in 3D modelling came from doing 3D animation in the 1990s when the hardware was very, very slow! Turning multi dimensional ideas into instructions for information and printing technology is what he does best and what he enjoys.

Frank lives an alternative off grid-in-a-village lifestyle. Alternative because there is no 9-5, and it pairs up-to-date technology with frugal-living. Not always an easy balance.

Frank lives and co-works with his multi-creative co-author and wife Angie who is a well known (in the small world of miniatures) craft and craft business writer. They share the home of too many cats.

## Thanks and acknowledgements

Tommy Hogg for constant support and encouragement of many kinds during my printing experiments.

Lisa Sones-Peck who has been my style consultant during the design of some of these projects.
The patrons without whom the past few years would have been much more difficult if not impossible and none of the last few books would have been written. You are all our supporting angels!

Christine McKechnie, Jacquie Hall, Maggie Dokic, Rachel Timmerman, Natalie Martin, Robyn Stewart Rachel C Taylor, Denise Pinnell, Gillian Mason-Thompson, Petra Sujiker and, of course, many more who didn't give permission to be named but who are just as important to me. Thank you all!

## Reviews

Reviews really help self-published authors compete in the field. Please review or rate this book if you have time. We really appreciate every single review.

Let us know if you have published a book on Kindle Unlimited and we will happily review your book if we like it.

## Patreon
## Why we love our patrons

If you've never heard of Patreon before, imagine you could be a 'patron of the arts' in some small way helping your favourite artists to continue working, inventing and teaching in their specialist area. Artists no matter how well known in their field often have no regular guaranteed income and often give away their inspiration for free because until now there wasn't an easy method to gain an income from day to day teaching, support and skill sharing.

This subscription service is an easy way to connect artist teachers with their students and followers, and as a way for the 'patrons' to give the level of support they are comfortable with and receive in return (sometimes personalised) perks such as early access to new ideas, live patron only videos and little samples of work to help you visualise stages of work and qualities of colour. As well as advance knowledge of really new ideas before they ever get to publication. Some ideas of how I made things which never even reach the books which I call my 'daft ideas' for example how I made the awning for the dollhouse shop on the front just using parts from an old umbrella! For benefit patrons I'm also able to send out little found 'things' which might inspire you, or samples of my new tools before they go into full production.

Many thanks to our current 70+ patrons some of who have been with me for several years now. You've all given me courage to start with new things like this book. The Patreon thing has really helped me because it's like having 70 sets of shoulders to lean on. 70 therapists and 70 special friends to share my daft ideas with and see if they work. Or at least are interesting enough for you not to walk away! 70 people who understand that no matter how well known an artist is they still may struggle from time to time. That's worth so much!

**www.patreon.com/angie_scarr**

---

### Making Miniature Food & Market Stalls

Angie's first book Published by Guild of Master Craftsman Publications. A bestselling introduction to making polymer clay miniature food. This is an updated edition.

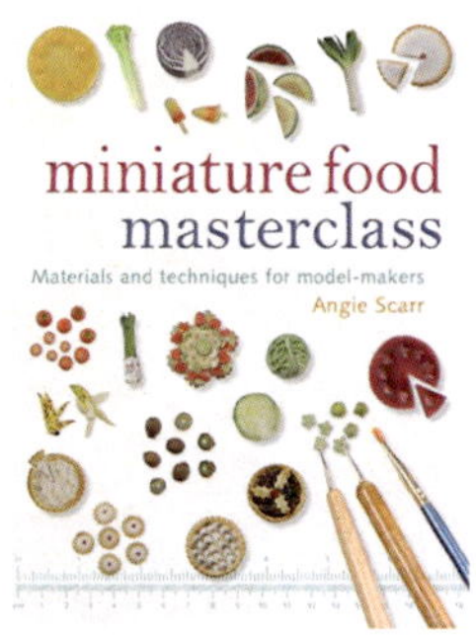


### Miniature Food Masterclass

Angie's second book with GMC. Also still a bestseller this one continues the journey of exploration into what polymer clay can replicate.

# Other books by Sliding Scale

## Making Miniature Flowers and Plants

The second in Angie's polymer clay stencilled flowers books. With new multilayer, 'wiring' and assembly techniques.

## The Dollhouse Flower Shop

The innovative idea of stencilling flowers in polymer clay/liquid clay mix. Using stencils available from Angie.

## Simple Mold Making

A book full of quick molding ideas for miniaturists, model makers, animators and jewellers using 2 part silicone mold material.

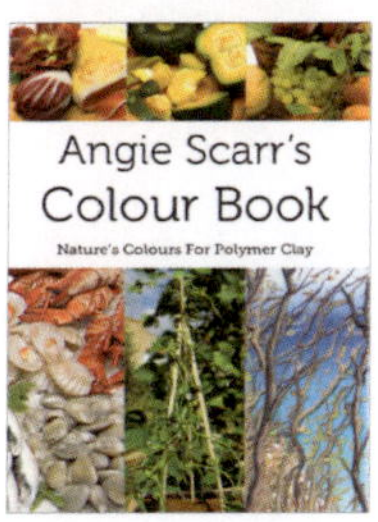


## Angie Scarr's Colour Book

New large edition book that asks the big questions about colour realism in polymer clay, helping you towards work that is so realistic it jumps out from the rest.

## The Miniature Gardens Book

Have you ever fancied making more than just a flower garden in miniature? Angie gives you several garden styles and lots of new ideas.

## Making Miniatures with 3D Printers

Go from nothing to printing miniatures in minutes. Fully hands-on from a simple vase to a full shop. All made with easily available free software.

## Cricut & Construct #1

Flowers and plants for cutting machines, from using our designs to the techniques for making your own.

## 3D Pen #1 Fairy Houses and Fantasy Gardens

A handy pattern book for anyone of any age who is looking for a project to make with their 3D pen. Excellent addition to a 3D pen gift.

## How to Self Publish Your Craft Book

Quick start guide to publishing your own books.

## Your Creative Business

Advice on all aspects of craft business from pricing and marketing through to multiple income streams.

## Your Creative Business Workbook

To go with the main book or as a stand alone.

---

## Making it Small- Biography

Angie never lived an 'ordinary life'. When she and Frank met it became less ordinary still. A story of the love of crafts, miniatures, self building and life in a small pueblo in Spain.

## The Frugalist

A look at revaluing your time, living better for less and gently preparing for unexpected crises. If life sometimes feels tough this book might just help.

**patreon.com/angie_scarr**
*Support our writing*

**facebook.com/angiescarr.miniatures**
*Our Facebook page where we let everyone know what is going on*

**instagram.com/angiescarr**
*Photos of work in progress*

**youtube.com/user/angiescarr**
*For tutorials, howtos and videos about crafts and miniatures*

**pinterest.com/angiescarr**
*Links to our work all over the internet*

**ko-fi.com/angiescarr**
*If you've learned something valuable buy Angie a coffee*

**tiktok.com/@angiescarr**
*Video shorts*

**@littleoldladywho1.bsky.social**
*Bluesky: the new Twitter*

---

# etsy.com/shop/AngieScarrCrafts

## Our Etsy digital store for plotter / cutter files

*Spiral staircases, market carts, flowers, leaves, boxes, flowershops and more*

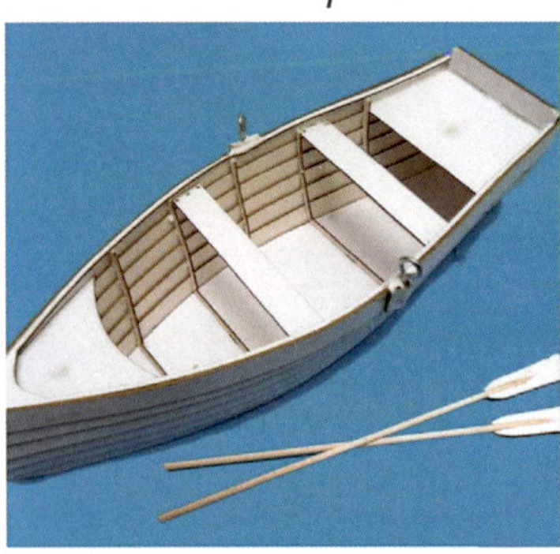

# etsy.com/shop/FranksCraftFiles new: 3D printer files

---

# angiescarr.com

## For moulds, stencils, kits, books, miniatures and other craft materials

*Delivery worldwide and trade terms available*

www.ingramcontent.com/pod-product-compliance
Lightning Source LLC
Chambersburg PA
CBRC091212220926
51847CB00007B/56

* 9 7 8 8 4 1 2 6 0 2 3 9 5 *